RUBBER PLANTATIONS AND CARBON MANAGEMENT

RUBBER PLANTATIONS AND CARBON MANAGEMENT

Arun Jyoti Nath, PhD
Biplab Brahma, PhD
Ashesh Kumar Das, PhD

Apple Academic Press Inc.
3333 Mistwell Crescent
Oakville, ON L6L 0A2, Canada

Apple Academic Press Inc.
1265 Goldenrod Circle NE
Palm Bay, Florida 32905, USA

© 2020 by Apple Academic Press, Inc.
Exclusive worldwide distribution by CRC Press, a member of Taylor & Francis Group
No claim to original U.S. Government works
International Standard Book Number-13: 978-1-77188-783-0 (Hardcover)
International Standard Book Number-13: 978-0-42902-349-1 (eBook)

All rights reserved. No part of this work may be reprinted or reproduced or utilized in any form or by any electric, mechanical or other means, now known or hereafter invented, including photocopying and recording, or in any information storage or retrieval system, without permission in writing from the publisher or its distributor, except in the case of brief excerpts or quotations for use in reviews or critical articles.

This book contains information obtained from authentic and highly regarded sources. Reprinted material is quoted with permission and sources are indicated. Copyright for individual articles remains with the authors as indicated. A wide variety of references are listed. Reasonable efforts have been made to publish reliable data and information, but the authors, editors, and the publisher cannot assume responsibility for the validity of all materials or the consequences of their use. The authors, editors, and the publisher have attempted to trace the copyright holders of all material reproduced in this publication and apologize to copyright holders if permission to publish in this form has not been obtained. If any copyright material has not been acknowledged, please write and let us know so we may rectify in any future reprint.

Trademark Notice: Registered trademark of products or corporate names are used only for explanation and identification without intent to infringe.

Library and Archives Canada Cataloguing in Publication

Title: Rubber plantations and carbon management / Arun Jyoti Nath, PhD, Biplab Brahma, PhD, Ashesh Kumar Das, PhD.
Names: Nath, Arun Jyoti, author. | Brahma, Biplab, author. | Das, Ashesh Kumar, author.
Description: Includes bibliographical references and index.
Identifiers: Canadiana (print) 20190134690 | Canadiana (ebook) 20190134747 | ISBN 9781771887830 (hardcover) | ISBN 9780429023491 (ebook)
Subjects: LCSH: Rubber plantations—Environmental aspects. | LCSH: Carbon dioxide mitigation.
Classification: LCC HD9161.A2 N38 2020 | DDC 338.4/76782—dc23

Library of Congress Cataloging-in-Publication Data

Names: Nath, Arun Jyoti, author. | Brahma, Biplab, author. | Das, Ashesh Kumar, author.
Title: Rubber plantations and carbon management / Arun Jyoti Nath, PhD, Biplab Brahma, PhD, Ashesh Kumar Das, PhD.
Description: Oakville, ON, Canada ; Palm Bay, Florida : Apple Academic Press, 2019. | Includes bibliographical references and index. | Summary: "With the increasing atmospheric carbon dioxide concentration and the resulting environmental consequences for plants, it is necessary to consider the future of rubber plantations, an important source of latex for rubber production. In Rubber Plantations and Carbon Management, the authors explore the ecology of rubber plantations in the context of carbon management under a scenario of our changing climate. The authors provide an in-depth study of the carbon stock and sequestration potentiality of rubber plantations. The volume also provides information on a biomass estimating model that can be used in future study of non-harvesting biomass estimation for a variety of plants. Topics in the volume address ecosystem biomass carbon losses through clear felling management systems, ecosystem carbon sequestration, and fine root dynamics in carbon balance under a rubber plantation management system. The volume concludes with an examination of future strategies and determines that rubber plantation development by replacing natural forests should not be encouraged, though progressive land use changes, for example, conversion of degraded forest lands or grasslands to rubber plantations, can enhance biomass and soil stock. The expansion of rubber tree plantations with best management practices in degraded forest, grasslands, or fallow lands can offer an ecological stability with significant socioeconomical support. This volume is the first of its kind on this important topic. The volume will be valuable for faculty and students, cultivators, researchers, rubber board officials, and many others. It will also aid in policy formulation at the scientific community level for those who are working on climate change adaptation and mitigation aspects"-- Provided by publisher.
Identifiers: LCCN 2019026709 (print) | LCCN 2019026710 (ebook) | ISBN 9781771887830 (hardback) | ISBN 9780429023491 (ebook)
Subjects: LCSH: Rubber plantations--Environmental aspects. | Carbon dioxide--Environmental aspects. | Carbon dioxide mitigation.
Classification: LCC HD9161.A2 N38 2019 (print) | LCC HD9161.A2 (ebook) | DDC 633.8/9520286--dc23
LC record available at https://lccn.loc.gov/2019026709
LC ebook record available at https://lccn.loc.gov/2019026710

Apple Academic Press also publishes its books in a variety of electronic formats. Some content that appears in print may not be available in electronic format. For information about Apple Academic Press products, visit our website at **www.appleacademicpress.com** and the CRC Press website at **www.crcpress.com**

About the Authors

Arun Jyoti Nath, PhD
Arun Jyoti Nath is Assistant Professor in the Department of Ecology and Environmental Science, Assam University, Silchar, Assam, India. He is an ecologist with diverse experience in research and teaching. His research focuses on biocarbon sequestration. Dr. Nath has obtained advanced training on carbon sequestration research from the Institute of Botany and Ecophysiology, Szent Istvan University, Godollo, Hungary; the National Bureau of Soil Science and Land Use Planning, Nagpur, India; and the Carbon Management and Sequestration Center, Ohio State University, USA. Dr. Nath is an academic editor for the journal *PLOS ONE*.

Biplab Brahma, PhD
Biplab Brahma is a senior research scholar in the Department of Ecology and Environmental Science, Assam University, Silchar, Assam, India. His research interests include ecosystem carbon management in terrestrial ecosystems.

Ashesh Kumar Das, PhD
Ashesh Kumar Das is Professor in the Department of Ecology and Environmental Science, Assam University, Silchar, Assam, India. His research interests include tree diversity and ecology, carbon management in forest and agricultural ecosystem, and soil ecology. Prof. Das has published more than 120 research papers in leading national and international journals.

Contents

Abbreviations .... *ix*
Preface .... *xi*

1. General Introduction 1
2. Biomass Models and Biomass Carbon Stocks 19
3. Potential Loss of Biomass Carbon 43
4. Impact of Land Use Changes on Soil Organic Carbon 61
5. Ecosystem Carbon Sequestration 83
6. Soil Carbon Sequestration 107
7. Fine Root Dynamics and Carbon Management 121
8. Conclusions and Recommendations 143

Index *147*

Abbreviations

AC	active carbon
AGB	aboveground biomass
AIC	Akaike information criterion
Arp	Areca (*Areca catechu*) plantation
BD	bulk density
BGB	belowground biomass
CDM	clean development mechanism
CGS	combating greenhouse gas
CI	confidence intervals
CL	confidence limits
DBH	diameter at breast height
DF	disturbed forest
DW	dry weight
ELRBP	estimated live root biomass production
FRP	fine root production
GHG	greenhouse gas
HSD	honestly significant difference
IG	Imperata grassland
KP	Kyoto Protocol
LC	labile C
LLC	less-labile C
LUC	land use change
MAPE	mean absolute percentage error
MODR	monthly observed dead root biomass
MOLR	monthly observed live root biomass
MRT	mean residence time
NEI	northeast India
NF	natural forest
NLC	non-labile C
PB	pan (*Piper betle*) jhum (slash and mulching) agroforestry
PC	passive carbon
PDW	part dry weight
RDW	root dry weight

REDD+	Reducing Emissions from Deforestation and Forest Degradation
RMSE	root mean square of error
RP	rubber (*Hevea brasiliensis*) plantation
SOC	soil organic carbon
SOM	soil organic matter
StD	stand density
TOC	total organic carbon
VLC	very labile C

Preface

The atmospheric carbon dioxide (CO_2) concentration has reached the highest-ever level of 410 ppm in 2017. This increase in CO_2 concentration is mainly caused by fossil fuel burning, cement industries, and land-use changes. Future prediction indicates that by 2050, CO_2 concentration will be around 500 ppm, which will lead to the extinction of several plant and animal species. Therefore, to maintain the atmospheric CO_2 concentration to a considerable level, different international protocols/organizations have been adopted/developed. The Kyoto Protocol (KP), adopted in 1997, is one of the most significant protocols adopted to maintain the atmospheric CO_2 concentration. Thereafter, several conventions and treaties have been developed under KP to manage the existing carbon pools and to increase their carbon sink potentiality as well. However, to fulfill the needs of the exponentially increasing population of the world, the forests are under serious threat. Plantation systems in this regard help in reducing the pressure on natural forests.

Rubber plantations are one of the rapidly growing plantation systems across the tropical world due to their significant commercial value and demand. The climate of the tropical region is suitable for its growth, and in Southeast Asia annual increment of rubber plantations is accounted to be 1.3%. This high pace of development of rubber plantations has caused significant changes in the native land use in Southeast Asia. Rubber plantations in Southeast Asia have been developed either by clearing the native forest (retrogressive land-use change) or by developing plantations on degraded forests or grasslands (progressive land-use change). Tropical forests are known to be the "lungs of the earth" due to their significant contribution to the global carbon cycle. Therefore, the changes of native land use for any monoculture plantation system may accelerate a climate-change phenomenon. On a global scale, ~11 M ha land area is currently under rubber plantation, which has been attributed for ecosystem carbon loss, disruption of hydrological and nutrient cycle, and loss of biodiversity. With these environmental consequences, it is necessary to explore the potentiality of rubber plantations in environment management.

In this book, the authors explore the ecology of rubber plantations in environment management. The present book is the first such comprehensive work that explores the ecology of rubber plantation in carbon management under a scenario of changing climate. In this contribution, the authors explore the carbon stock and sequestration potentiality of rubber plantations. This book also provides information on the biomass-estimating model that can be used in future study of nonharvesting biomass estimation. The present contribution also addresses ecosystem biomass carbon losses through clear felling management systems, ecosystem carbon sequestration, and fine root dynamics in carbon balance under the rubber plantation management system. The present contribution concludes with remarks that the development of rubber plantations by replacing natural forests should not be encouraged; however, progressive land-use changes, for example, conversion of degraded forest lands or grasslands to rubber plantations, can enhance biomass and soil organic carbon stock. Adoption of rubber tree plantations with best management practices, for example, rubber agroforestry in degraded forests or grasslands can offer an ecological stability with significant socioeconomical support.

CHAPTER 1

General Introduction

1.1 DISTRIBUTION OF RUBBER PLANTATIONS

The rubber tree, *Hevea brasiliensis* (Willd. ex A. Juss.) Müll. Arg., is a tropical plant that grows in the region with a temperature range of 20–30°C and an average annual precipitation of >1500 mm (Phuc and Nghi, 2014). Rubber trees also grow up to 600 m elevations and are distributed in the 10° latitudinal regions of either sides of equator (Balsiger, 2000). Rubber is a "hot commodity" with consumption increasing worldwide at an average rate of 5.8% per year since 1900 (The Rubber Board, 2005). Synthetic rubber accounts for approximately 57% of the total, but natural rubber is cheaper and of superior quality for high-stress purposes. Jet and truck tires are almost entirely made of natural rubber. Prachaya (2009) forecasts that total rubber consumption will increase from 22.1 million ton in 2008 to 23.2 million tons by 2018 and the relative share of natural rubber will increase from 43% in 2008 to 48% by 2018. Consumption of natural rubber is anticipated to increase from 9.6 million tons in 2008 to 13.8 million tons by 2018—a growth of 3.7% per year (Prachaya, 2009). In order to fulfill the future demand of natural rubber consumption, the area under rubber tree plantations are growing significantly throughout the world, most specifically in Tropical Asian countries. In mainland of natural rubber production includes portions of Thailand, Indonesia, Myanmar, India, Vietnam, and China (Figs. 1.1 and 1.2; Rubber Statistical News, 2012). It has been estimated that only Asia accounts for 97% of the world's natural rubber supply, with the vast majority coming from Thailand (31%), Indonesia (27%), and Malaysia (9%), India (8%) (Rubber Statistical News, 2012).

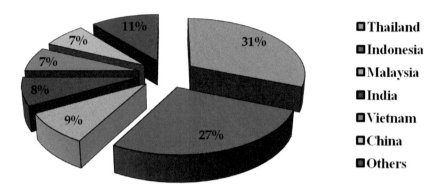

FIGURE 1.1 (See color insert.) Share of natural rubber production in 2011.

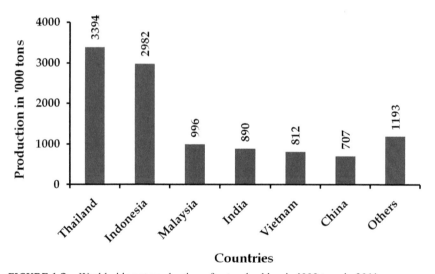

FIGURE 1.2 Worldwide net production of natural rubber in '000 tons in 2011.

1.2 GLOBAL DISTRIBUTIONS AND PRODUCTION

Rubber is native to the Amazon basin and it has been historically cropped in the equatorial zone between 10°N and 10°S in areas with twelve months of rainfall. This region includes portions of southern Thailand, south eastern Vietnam, and southern Myanmar. In the early 1950s, in attempt to secure economic development, China invested heavily in research on

General Introduction

growing rubber in environments perceived to be marginal in terms of cooler temperatures and having a distinct dry season. State rubber plantations were established in Hainan and Yunnan provinces in areas that lie as far north as 22° north latitude. China's success in growing rubber in these "non-traditional" environments greatly expanded the habitat in which rubber is planted. More recently, rubber plantations have expanded by more than 1,000,000 ha in upland areas of China, Laos, Thailand, Vietnam, Cambodia, and Myanmar where rubber trees were not traditionally planted (Li and Fox, 2012; Ziegler et al., 2009; Manivong and Cramb, 2008). Plantation forests expanded dramatically, displacing natural forests and shrubs/grasslands primarily at low- and mid-elevations in Hainan's natural forests (Zhai et al., 2012). Although Indonesia occupies the second position in natural rubber production, it had the highest area under rubber tree plantations of 3,435,000 ha by end of the 2009, that has increased to 3,621,000 ha in 2016 based on the assessment of Indonesian Rubber Association. However, Thailand, Malaysia, China, and India, respectively occupy the second, third, fourth, and fifth positions in occupying area under rubber tree plantations (Fig. 1.3; The Rubber Board, 2005).

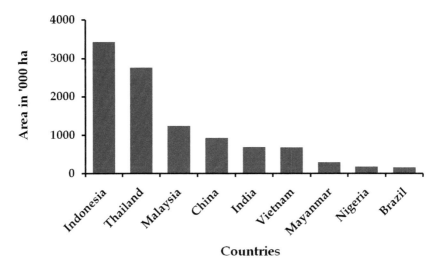

FIGURE 1.3 Worldwide net area under natural rubber tree plantations in '000 ha in 2009. The figure illustrates the area under rubber tree plantations of China and Myanmar in 2008 and 2006, respectively.

Distribution of rubber plantations are largely affected by the geographical constrains, because of its meteorological dependency in growth, specifically on average temperature and rainfall of the region (Liu et al., 2013). In Indonesia the maximum share of land under rubber plantation is distributed in Sumatra and Kalimantan, which has been converted from natural forests (Lambert and Collar, 2002). Moreover, replacing coffee, cocoa, and tea by the rubber plantation, due to its vast profitability, is common in other regions like Riau and Jambi. Smallholder cultivators of Sumatra and Kalimantan are playing a significant role in the dramatic increment of the rubber plantations in the country. Around 84% of the total area under rubber plantations of Indonesia in 2015 was contributed by the smallholders of the country (Indonesian Rubber Association). Although Indonesia is the leader of area under rubber plantations, the total production of Indonesia is at the second position in the world. According to the estimation of Association of Natural Rubber Producing countries (ANRPC) the total annual production of natural rubber in Indonesia was around 3.2 million tons in 2014 (ANRPC, 2015). Out of the total country production only 15% has been used domestically and the rest around 85% of the total production of the country has been exported to the other Asian countries followed by North America and Europe. However, Thailand takes the first position in natural rubber production. Around 4.5 million tons of natural rubber has been produced in Thailand in 2016, out of which 3.5 million ha of land area of the country. Out of the total country production, Thailand exports 3.6 million tons of natural rubber globally. The climatic suitability in Thailand encourages smallholders to grow rubber plantations in South and Northeastern regions. At least 95% of its total cultivation of the country has been contributed by the smallholders in Thailand. Rubber plantations in Thailand are largely distributed in South and Northeastern region contributing 67% and 17% of the total area, respectively (Office of Agricultural Economics, 2010). Although Malaysia occupies the third position in area under rubber plantation, in terms of production, it lies in sixth position after China. Vietnam produces around 1 million tons of natural rubber in 2013 and took over the third position after Indonesia. However, Thailand, Indonesia, and Vietnam together in 2013 produced around 73% of the total natural rubber production of the world. India produced around 0.9 million tons of natural rubber during the same year. Climates of southern states and northeastern states of India is suitable for the growth of rubber plantations results almost the all rubber plantations of the country grows in these regions only. Kerala produces around 90% of the total natural rubber production of the country and covers around 0.5 million ha of land area followed by Northeastern States covering around 0.15 million ha land (The Rubber Board, 2011). China is the largest consumer of the world consuming at least around 36% of the global consumption and will increased up to 39% of the global. It is the largest importer of natural rubber in the world estimating around 40% of the total global market. With this increasing demand of natural rubber, the consumption over the world will be increased by at least 4.6% during 2017–2022 (Natural Rubber Trends and Statistics, 2018).

1.3 WORLD CARBON BUDGET: PAST, PRESENT, AND FUTURE

The greenhouse gas (GHG) emission in the atmosphere has increased substantially to achieve the average carbon concentration at 406 ppm (1 ppm = 1 part per million) in 2016 from 277 ppm in 1750 mainly due to anthropogenic activities (Lal, 2008; Le Quere et al., 2016). Cumulative anthropogenic CO_2 (carbon dioxide) emission during 1750–2011 in the atmosphere was 555 Gt C (1 gigaton = 1 billion tons), of which 375 Gt C was emitted from fossil fuel burning and cement industries and another 180 Gt C from the deforestation and other land use changes (IPCC, 2013). In contrast, for the last decade (2005–2014) the annual CO_2 concentration was increased by 4.4 Gt C year^{-1}, that was chiefly shared by the fossil fuel burning and industries and land use changes by 9 and 0.9 Gt C year^{-1} respectively, that was accounted by 91% and 9% respectively to the total emission (Le Quere et al., 2015). However, over the period of 1970 to 2014, the biggest source of CO_2 emission from fossil fuel burning was increased from 4.7–9.8 Gt C year^{-1} (Le Quere et al., 2015). The annual average emission from land use change was observed to be decreased from 1.3 to 1.1 Gt C year^{-1} over the same time frame (Le Quere et al., 2015). The projected atmospheric concentration of CO_2 is 500 ppm by 2050, resulted the temperature increase and sea level rise (Ciais et al., 2013). These irreversible changes may consequently extinct of many species from the earth with changes of many landscapes (IPCC, 2007).

Therefore, in order to combat their reversible consequences of climate change it became crucial to trim down the increasing trend of CO_2

concentration through reducing sources or enhancing sinks. Reducing the emission of all the GHG gasses during 2013–2020 up to a minimum level of 18% of their 1991 level was decided by the parties in the Doha amendment to the Kyoto Protocol (KP; Doha Amendment to the KP, 2012). To achieve the goal of stabilizing atmospheric CO_2 in a cost effective way, the KP establishes different mechanisms that can be adopted by the participating countries (UNFCCC, 1997). However, a number of global facilities including Reducing Emissions from Deforestation and Forest Degradation (REDD+), Clean Development Mechanism (CDM), Green Climate Fund have been developed and implemented so far (UNFCCC, 2011, 2008; IPCC, 2007). CDM is one of the three mechanisms under flexible market mechanisms adopted in the KP. This mechanism encourages the role of biomass in carbon stock management through afforestation and reforestation activities. The growing political (the Kyoto Protocol) and economical interest in carbon accounting requires ways and means of estimating carbon sources and sinks.

Carbon sinks are associated with CO_2 absorption by the oceans (2.4 Gt C year^{-1}; Plattner et al., 2002) and terrestrial uptake (3.6 Gt C year^{-1}; Houghton and Goodale, 2004; Prentice et al., 2001). Global (151–108 ha terrestrial storage area) carbon stocks in terrestrial vegetation (466 Gt C) and in the soil down to a depth of 1 m (2011 Gt C) appear quite limited compared to the 39,000 Gt C in oceans and the 75–106 Gt C in rocks but are of the same order of magnitude as those stored in the atmosphere (760 Gt C; Bolin et al., 2000). However, the global soil organic carbon (SOC) stock of ~1.5×10^3 Pg (Petagram; 1 Pg = 10^{15} g) has been accounted as two and three folds higher than that of atmosphere and vegetation respectively (Jobbagy and Jackson, 2000). This major pool of carbon (C) has significant effect on atmospheric C concentration by releasing and sequestering processes. However, the distribution of C in vegetative pool or soil pool in terrestrial landscapes varies on land use and land cover characteristics.

1.4 TROPICAL FOREST AND CARBON

Tropical forests are the major source/sink of C (Wei et al., 2014) due to their strong (46%) impact on the global terrestrial C cycle (Bloom et al., 2016). Since the C in tropical forests are stored in different pools [vegetation (67%) and soil (33%)] and their accumulation rate and sink capacity vary with climatic variations, vegetative cover and intensity of human interventions (Dung et al., 2016; Ngo et al., 2013; Lal, 2008; Lawrence et al., 2007). A small change in tropical forests may strongly perturb the global C cycle while also jeopardizing several ecosystem services (Polasky et al., 2011). The land use change from forests to any land use degrades native vegetative structure and lowers the ground water level (Ahrendset al., 2015), which further affect the ecosystem's C dynamics (De Camargo et al., 1999). Since the forestry and agroforestry systems possesses significant ability in C stock and sequestration potentiality (Kirby and Potvin, 2007; Sharrow and Ismail, 2004; Roshetko et al., 2002; Sanchez, 2000); sustainable forest management is the most supportive strategy to remove C from the atmosphere (UNFCCC, 2008). However, to accomplish the demands of increasing population needs the world's natural forests are under serious threat. Therefore, plantation forestry and agroforestry are the reliable option to switch the target from natural forest (FAO, 2010).

Plantation forests constitute 264 million ha (M ha) globally and represents 6.4% of the world's forest land (FAO, 2010). Since, properly managed forest plantations could provide numerous social values and can be an important measure for climate change mitigation (UNFCCC, 2008), it attracted the interests of policy makers in developing policies related to removal of atmospheric CO_2 through natural C sink (Lorenz and Lal, 2014). Worldwide a net uptake of 1.1 to 1.6 Pg CO_2 per year could be achieved by developing plantation forestry or agroforestry as reforestation and afforestation activities between 1995 and 2050 (IPCC, 2003). Therefore, accurate estimation and management of biomass and soil carbon stocks under plantation development activities could pave the ways for developing strategy to mitigate the atmospheric CO_2 concentration level. Moreover, the vegetative carbon pools comprises above and below ground biomass of trees.

The accurate estimation of total above and below ground biomass carbon stock could help to development of the inventory dataset. Tree biomass can be estimated through harvest and non-harvest methods (Brown, 1997). Available generalized models are useful tools for estimating plant biomass of forest, agroforestry and plantation forestry non-destructively (Chave et al., 2001; Houghton et al., 2001; Brown et al., 1989; Cunia, 1987; Crow, 1978). However, due to variations in plant architectures and complexity in associations of trees, the biomass estimation through generalized models is fraught with errors (Brown and Schroeder, 1999; De Oliveira and Mori, 1999; Shepashenko et al., 1998; Tar-Mikaelian and Korzukhin, 1997). Generalized models have been used for biomass estimation without consideration of the tree age or plantation age. Tree age is a vital factor that influences biomass accumulation (Fatemi et al., 2011), suggesting the need for age-specific biomass models for precise estimation. Therefore, it is a vital need for scientific discussions on the usefulness of generalized models specifically for biomass estimation of mono-specific tree plantations for future policy making.

The SOC concentration or distribution of SOC varies with the vertical and horizontal variances. It varies with the land use characteristics, management practices and the soil depth. Since, SOC in soil is the translocation form of litter C, therefore, upper soil contain higher C concentration with high microbial activity than the lower soil depths. However, management of SOC stock up to the soil depth of 0–40 cm has been widely recommended during COP-21 in Paris-2015 to achieve the goal of food security and climate change mitigation. Moreover, depending on the residence time of organic carbon in soil, it has been categorized as active and recalcitrant or passive pools (Chan et al., 2001). The microbial and derivative products which are very active in recycling with in a period of 5 years have been recognized as the active pool (AC). The actively recycling AC pools are the chief source of soil nutrients (Chan et al., 2001). Therefore, the productivity and quality of soil are greatly influenced by the AC pool fractions (Mandal et al., 2008; Chan et al., 2001). However, due to its low residence time the C sequestration potentiality of AC pools is very low (Luo et al., 2003). However, the passive pool (PC) which possesses relatively longer residence time can act as a significant contributor to the total organic carbon stock (Mandal et al., 2008). Therefore, PC pools can act as a reliable indicator of C sequestration potentiality of a system (Paul et al., 2001).

1.5 LAND USE CHANGE AND EXPANSION OF RUBBER PLANTATION

1.5.1 The Primitive Agriculture

The only human habitat of the Universe has become a "Used Planet" during last several centuries due to unrestrained human driven activities (Ellis et al., 2012). Changes of natural land architectures according to the human needs have been witnessed since prehistorical period. Around 10,000 years ago when the ice age was prevailed in the earth, humans were dedicatedly involved with hunting of animals to survive (Martin and Sauerborn, 2013). Toward the end of the ice age the climate of the earth has begun to change with rise in temperature and the rainfall intensity. The vegetative characteristics of the Earth also began to change resulting into disappearance of numerous existed species. Introduction of agriculture to produce an alternative source of food during the last centuries of ice ages were mainly owing to the global climate change and declining wildlife population (Martin and Sauerborn, 2013). At the beginning only a few area were dedicated for agriculture to produce seed and plants (Lamb, 1982; Jager and Barry, 1990). Cultivated plant and seeds were served as the alternative food in case of failure in haunting (Martin and Sauerborn, 2013). Gradually due to lacking of wildlife animals, the dependency on alternative food source has started to gain much attention of the primitive people.

At the beginning, the selection of cultivable seed and plants were believed to be unconscious or accidentally. Lost or unconscious spread of seeds near by the resting and habitat places and harvesting of these regenerated plant species for food has been supported by the theories of agriculture development (Martin and Sauerborn, 2013). The preferred characteristics as food source of these harvested plants were appeared at the early stage than that was seen in wild habitat. The purposive selection of seed or plants for further food production was initiated at this stage of agriculture development. The theory of domestication involves the purposive selection of plants and animals over the centuries eventually made their genetic modification toward more suitable for human desires (Martin and Sauerborn, 2013), called domesticated crops.

Eleven regions around the world are known to be the origin of agriculture of domesticated crops. It is known that around 11,000 years back the agriculture was originated in the zone of "Fertile Crescent" covering Jordan, Syria and the valleys of Euphrates and Tigris river. Cultivation of crops in other region was introduced from the centre of origin. The wheat species emmer (*Triticum dioccum*) and einkorn (*Triticum monococcum*) were known as the first domesticated crops that were cultivated as agricultural crop in "Fertile Crescent." Other crops like barley (*Hordeum vulgare*), pea (*Pisum sativum*), lentil (*Lens esculenta*), and chickpea (*Cicer arietinum*) were started to cultivate in Fertile Crescent. Afterwards around 1000 years later, that is, around 10,000 years back of the present day, rice (*Oryza sativa*) was started to cultivate in tropical and subtropical Southern China and Middle East. However, the highlands of tropical Southeast

Asian countries were experienced with banana (*Musa sp.*) and sugar cane (*Saccharum officinarum*; Denham et al. 2003), when the agriculture in South America was began with (*Cucurbita sp.*; Piperno and Stothert 2003). In North America sunflower (*Helianthus annus*) was started to cultivate as an oil crop around 4000 years back from the present day.

Although the primitive agriculture was started in the uphills and valleys of the mountainous region but afterwards during the Greek era the agriculture was introduced in the plains. According to archaeological and paleological studies, overgrazing and farming activities on the valley slopes resulted catastrophic soil erosion in the Southern Argolid, the Argive Plain and the Larissa Basin of Greece during 6000–2500 years before the present. Greek philosopher Aristotle and Plato described the destruction of the environment around 12,000 years before the present was due to over exploitation of primary forests and that caused great alteration of environment. The wood used in the development of great Athenian fleet of ships signifies the over exploitation of forests of Greece. Utilization of forest trees as fuel wood to smelting of iron ores during seventh century B.C. in the Lebanon was also considered as a factor behind forest clearance of Greece. However, during the Roman era forests were being cleared in order to supply the fuel wood. Since more than 90% of the forest cleared trees were used as fuel wood, the other purpose like building materials and materials for war were also exists. The same fate was written in the history of the forests of China and Central America later.

By the advancement of civilization and human population the requirements have increased day by day and are still continuing. The world human population has increased to 7.6 billion in 2018 from 0.005 billion on average in 10,000 years back from the present day (McEvedy and Jones, 1978; LiviBacci, 2001; UN 2018). In order to maintain the food security with the exponential population growth, the trend of pioneer agriculture practices have been modified from no-till to tillage and no or less organic fertilizer to inorganic fertilizers. Moreover, technological advancement and modified food production systems were evolved to fulfill the requirement of rapid increasing world population. Over the period land use management practices also underwent changes from its native system to more human desired systems. In other words, the native ecosystems were modified according to the need of the human beings. It is known that during last 300 centuries the land use change has been altered dramatically due to technological advancement and industrial revolutions. During

1700, the land use activities were largely confined to Old World which was also known as Afro-Eurasia (Africa, Europe, and Asia collectively). Later, the land use pattern of the rest of the World also followed the pattern of the Old World. During 19th century, the relatively advanced technologies of Old World have been used for the modification of New World (America and Oceania). As a result, countries like the United States, Canada, Argentina, the Former Soviet Union, and Australia were established against the cost of huge natural forest covers. Exploitation of natural vegetation during 20th century has been largely confined in developing countries. In order to fulfill the requirements of the unregulated population explosion, the forests of these countries were being cleared for agricultural production. Since, the native land uses were primitively started to change for the fulfillment of food requirement, but the aspiration of the current land use change has become multidimensional. The 19th and 20th centuries are largely devoted for the development and expansion of plantation. With a significant monetary benefit the local peoples were largely encouraged to clear their forest for further establishment of cash crops in Malaysia, India, Africa, Brazil, China, and Central America. At a cost of establishment of sugar in Caribbean, cotton in North America, rubber in Malaysia, tea in India, coffee in Brazil, and bananas in Central America, a significant share of their natural vegetation has been lost. Establishment of different croplands in the native forests or grasslands is around 9.4 million km^2 over the world (Hurtt et al. 2011). However, around 40% of the total ice free land cover area has been changed into agriculture from natural vegetation (Ramankutty et al., 2008; Ellis et al., 2010). Specifically, 131 million ha area under forest/wetlands of the 13 nations of Asia (India, Sri Lanka, Bangladesh, Myanmar, Thailand, Laos, Cambodia, Vietnam, Malaysia, Brunei, Singapore, Indonesia and the Philippines) has been converted into other land uses during 1880–1980. However, over the century 106 million ha area has been accounted as increased under cultivated land. In Asia, 81% of the natural forest/wetlands have been converted into cultivated land during 1880–1980 (Richards and Flint, 1994). In Russia, the expansion of cultivated land during this century was estimated at 147 million ha. Whereas, during 1850–1985 around 370 million ha of forest lands have been converted into other land uses in Latin America. Out of this 35% land area has been devoted to either shifting cultivation or croplands. Introduction of sugar, rubber, and coffee in Brazil and Argentina triggered the loss of natural vegetation during 19th century in Latin America. Loss of natural forest from the Africa has altered by the expansion of cacao and coffee during early twentieth century.

1.5.2 Evolution of Rubber Plantations

The South America (the Brazilian and Bolivian region covering Amazon and Orinoco river basins) originated tree known as rubber tree or para rubber tree (*Hevea brasiliensis*) under Euphorbiaceae family is well known for its milky sap or latex. The use of this latex was started during 15th to 16th centuries by the Mesoamerica. Mesoamerica is one of the six places of the world where civilization was developed independently. The Olmec culture is one of the earliest culture of the world prevailed in Mesoamerica was evidently support for their use of natural latex for making balls for the Mesoamerican ballgame. Afterward the Maya and Aztec civilization were fond of with the use of natural rubber for making their containers and textiles. Thereafter in 1770, the English chemist Joseph Priestley introduced the use of rubber for rubbing the pencil marks on paper. Discovery of vulcanization in 1839 by Charles Goodyear encouraged the use of rubber latex to the global people. The first transportation of rubber seeds from Brazil to England was done by the Henry Wickham in 1876. The germinated seedlings were further transported to India, Sri Lanka, Indonesia, Malaysia, and Singapore. After globalizing the rubber cultivation, Malaysia became the most natural rubber latex producer country over the world. The use of rubber has boosted up after the invention of present form of tire by Philip Strauss in 1911 and accelerated commercial cultivation of rubber around the world. By the end of first quarter of the 20th century rubber latex was being started to use in door and window profiles, belts, gaskets, matting, and flooring.

Rubber tree is known as a tropical crop, prefer to grow in the climate prevailing 10° either sides of tropic of cancer. The ideal climate for growing rubber tree is with an annual mean temperature range of 20–30°C, precipitation >1500 mm, light to heavy clay with good water drainage (Phuc and Nghi, 2014). Due to the increase in demand and its high economic profitability rubber tree has become a boom crop across the tropical regions (Balsiger et al., 2000) and further extended its cultivation up to 20° latitudinal distances from the tropic of cancer in China. In order to eradicate the poverty the local government of Asian countries started to encourage people for rubber tree cultivation in their private lands and prioritized for secondary and degraded forest reclamations (Balsiger et al., 2000; Fox et al., 2013). As a result, the estate crops soon became a small holder crop in Asian countries and contributed about 75% of the total production of the region. With the increase in consumption of rubber over the years it has been projected the production of latex to be reach up to 23.2 million ton by 2020 from 22 million ton in 2008 (Prachaya, 2009; Natural Rubber Statistics, 2014). In order to meet this demand, natural rubber cultivation around the tropical region is rapidly increasing. The estimated area under the rubber plantation around the world has been increased at the rate of 1.71% per annum from 3.38 million ha in 1961 to 11 M ha in 2006 (FAO, 2010). More specifically, the valley and uplands of South East Asia rubber plantation expanded by ~9 M ha only within the last several decades (Manivong and Cramb, 2008; Ziegler et al., 2009; Li and Fox, 2011). Rubber plantation developments in Asia are mainly resulted of land use change from natural forest and grassland. It has been estimated that around 1 million ha land has been recently converted into rubber plantations in the uplands of China, Laos, Thailand, Vietnam, Cambodia, and Myanmar where rubber plantations were not traditionally grown (Manivong and Cramb 2008; Li and Fox 2011; Ziegler et al. 2009). The future perception of rubber plantation distribution could quadruple by 2050 replacing the evergreen broadleaf trees and Swidden related secondary forests (Fox et al., 2012). Specifically in Xishuangbanna of Southern China the lands under rubber plantations has increased by 175% between 2002 and 2010. In Xishuangbanna, the rubber plantation developed at a cost of natural vegetations at a significant level, resulted decline of natural vegetative cover from 49% in 1988 to 28% in 2006 of the total land cover (Li et al., 2007; Xu et al., 2014; Hu et al., 2008). However, 70% of the total rubber production of the world has been contributed by Indonesia, Thailand and Malaysia. Indonesia holds the highest lands under rubber plantations followed by Thailand and Malaysia. The idle climatic conditions of Indonesia encourage small growers throughout the country to develop plantation contributing 75% of the total production (BPS, 1999). In Indonesia, at least ten provinces from Aceh to Papua covering around 3.64 M ha land area of the country under rubber plantations and among them the South Sumatra province covering the highest area under rubber plantation around 0.81 M ha, accounts 23% of the total area under rubber plantations of the country. The expansion of rubber plantation in Indonesia is a result of land use conversion from natural forests (Vincent and Hadi, 1993). The reduction of natural forest area in Indonesia is the second highest in the world is around 0.8 M ha per year, which accounts around 19.7 M ha forest loss during the period of 1990 to 2013 (Directorate of Forest Resource Inventory and Monitoring, 2015). Thailand contributes the second highest land under rubber plantations in the world covering 3.5 M ha lands by 2014 throughout the country (Thailand's Rubber Industry, 2014). About 36% of the world rubber production has been contributed by Thailand by 2003 (Prabhakaran, 2010).

India is the fourth natural rubber producing country in the world, occupying the fifth position in distribution covering 711 thousand ha land by 2009 (The Rubber Board, 2005). Traditional and non-traditional are two rubber dominating regions divided upon the agro-climatic conditions of the country. Traditionally rubber trees are growing in the hinterlands of southwest coast, Kanyakumari of Tamil Nadu and Kerala (Ray et al., 2014; The Rubber board, 2005). The traditional regions of India contribute 90% of its total natural rubber production. Whereas, only Kerala contributes 89% of the countries production, covering 534,228 ha of land area followed by Tamil Nadu (Ray et al., 2014; The Rubber Board, 2005). The hinterlands of coastal Karnataka, Goa, Maharashtra, Andhra Pradesh, Orissa, Andaman and Nicobar Islands, and northeastern states are comprises for non-traditional rubber tree cultivation. Among the nontraditionally cultivated regions, Tripura occupying the first position in land area covering rubber tree plantations of 59,285 ha by 2011 (Rubber Statistical News, 2012; The Rubber Board, 2005). However, Karnataka, Assam, and Meghalaya occupy the second, third, and fourth positions in nontraditional cultivation of rubber trees.

KEYWORDS

- **monoculture plantation**
- **land use change**
- **Southeast Asia**
- **environment management**
- **rubber agroforestry**

REFERENCES

Ahrends, A.; Hollingsworth, P. M.; Ziegler, A. D.; Fox, J. M.; Chen, H.; Su, Y.; Xu, J. Current Trends of Rubber Plantation Expansions May Threaten Biodiversity and Livelihoods. *Glob. Environ. Chang.* **2015**, *34*, 48–58.

Balsiger, J.; Bahdon, J.; Whiteman, A. *The utilization, Processing and Demand for Rubber Wood as a Source of Wood Supply;* Forestry Policy and Planning Division, Rome, 2000.

Bloom, A. A.; Exbrayat, J. F.; Velde, I. R. V. D.; Feng, L.; Williams, M. The Decadal State of the Terrestrial Carbon Cycle: Global Retrievals of Terrestrial Carbon Allocation, Pools, and Residence Times. *PNAS* **2016**, *113*, 1285–1290.

Bolin, B. et al. Global perspective. In: Watson, R. T. et al. (Eds.). *Land Use, Land-Use Change, and Forestry: A Special Report of the Intergovernmental Panel on Climate Change*; Cambridge University Press: New York, 2000; pp. 23–51.

Brown, S. *Estimating Biomass and Biomass Change in Tropical Forest: a Premier UN FAO Forestry Paper;* FAO: Rome, 1997 (134); pp. 55.

Brown, S.; Schroeder, P. E. Spatial Patterns of Above Ground Production and Mortality of Woody Biomass for Eastern US Forests. *Ecol. Appl.* **1999**, *9*, 968–980.

Brown, S.; Gillespie, A. J. R; Lugo, A. E. Biomass Estimation Method for Tropical Forests with Applications to Forest Inventory Data. *For. Sci.* **1989**, *35*, 881–902.

Chave, J.; Riera, B.; Dubois, M. A. Estimation of Biomass in a Neotropical Forest of French Guiana: Spatial and Temporal Variability. *J. Trop. Ecol.* **2001**, *17*, 79–96.

Ciais, P.; Sabine, C.; Bala, G.; Bopp, L.; Brovkin, V.; Canadell, J.; Chhabra, A.; DeFries, R.; et al. Carbon and Other Biogeochemical Cycles. In *Climate Change 2013: The Physical Science Basis. Contribution of Working Group I to the Fifth Assessment Report of the Intergovernmental Panel on Climate Change*; Stocker, TF., Qin, D., Plattner, G. K. Tignor, M., Allen, S. K., Boschung, J., Nauels, A., Xia, Y., Bex, V., Midgley, P. M., Eds.; Cambridge University Press, Cambridge: United Kingdom and New York, NY, USA, 2013; pp. 465–570.

Crow, T. R. Common Regressions to Estimate Tree Biomass in Tropical Stands. *For. Sci.* **1978**, *24*, 110–114.

Cunia, T. Error of Forest Inventory Estimates: Its Main Components. *Can. J. For. Res.* **1987**, *17*, 436–441.

De Camargo, P.; Trumbore, S.; Martinelli, L.; Davidson, E.; Nepstad, D.; Victoria, R. Soil Carbon Dynamics in Regrowing Forest of Eastern Amazonia. *Glob. Chang. Biol.* **1999**, *5*, 693–702.

De Oliveira, A. A.; Mori, S. A. A Central Amazonian Terra Firm Forest I. High Tree Species Richness on Poor Soils. *Biodiversity Conserv.* **1999**, *8*, 1219–1244.

Doha Amendment to the Kyoto Protocol 2012. Doha, Qatar. http://unfccc.int/files/kyoto_protocol/application/pdf/kp_doha_amendment_english.pdf (accessed Jan 26, 2018).

Dung, L. V.; Tue, N. T.; Nhuan, M. T.; Omori, K. Carbon Storage in a Restored Mangrove Forest in Can Gio Mangrove Forest Park, Mekong Delta, Vietnam. *For. Ecol. Manag.* **2016**, *380*, 31–40.

FAO. Global Forest Resource Assessment, Food and Agricultural Organization of the United Nations (FAO): Rome, 2010.

Fatemi, F. R.; Yanai, R. D.; Hamburg, S. P.; Vadeboncoeur, M. A.; Arthur, M. A.; Briggs, R. D.; Levine, C. R. Allometric Equations for Young Northern Hardwoods: the Importance of Age-Specific Equations for Estimating Aboveground Biomass. *Can. J. For. Res.* **2011**, *891*, 881–891.

Houghton, R. A.; Goodale, C. L. *Effects Of Land-Use Change on the Carbon Balance of Terrestrial Ecosystems.* In *Ecosystems and Land-use Change*; DeFries, R. S., Asner, G. P., Houghton, R. A., Eds.; American Geophysical Union: Washington, DC. Geophysical Monograph Series, 2004; Vol. 153, pp. 85–98.

Houghton, R. A.; Lawrence, K. L.; Hackler, J. L.; Brown, S. 2001. The Spatial Distribution of Forest Biomass in the Brazilian Amazon: A Comparison of Estimates. *Glob. Chang. Biol.* **2001**, *7*, 731–746.

IPCC, Good Practice Guidance for Land Use. Land-Use Change and Forestry; IPCC National Greenhouse Gas Inventories Programme. Kanagawa, Japan, 2003.

IPCC, Summary for Policymakers. In Climate Change 2007: The Physical Basis, Contribution of Working Group I to the Fourth Assessment Report of the Intergovernmental Panel on Climate Change; Cambridge University Press, Cambridge: United Kingdom and New York, NY, USA, 2007.

IPCC, Intergovermental Panel on Climate Change Summary for Policymakers. In Climate Change 2013: The Physical Science Basis, Contribution of Working Group I to the Fifth Assessment Report of the Intergovernmental Panel on Climate Change; Cambridge University Press: Cambridge, United Kingdom and New York, NY, USA, 2013.

Kirby, K. R.; Potvin, C. *Variation in Carbon Storage Among Tree Species: Implications for the Management of a Small-Scale Carbon Sink Project. For. Ecol. Manag.* **2007**, *246*, 208–221.

Lal, R. Sequestration of Atmospheric CO_2 in Global Carbon Pools. *Energy Environ. Sci.* **2008**, *1*, 86–100.

Lawrence, D.; D'Odorico, P.; Diekmann, L.; DeLonge, M.; Das, R.; Eaton, J. Ecological Feedbacks Following Deforestation Create the Potential for a Catastrophic Ecosystem Shift in Tropical Dry Forest. *PNAS* **2007**, *104*, 20696–20701.

Le Quéré, C.; Moriarty, R.; Andrew, R. M.; Canadell, J. G.; Sitch, S.; Korsbakken, J. I.; et al. Global Carbon Budget 2015. *Earth Syst. Sci. Data* **2015**, *7*, 396.

Le Quéré, C.; Andrew, R. M.; Canadell, J. G.; Sitch, S.; Korsbakken, J. I.; Peters, G. P.; et al. Global Carbon Budget 2016. *Earth Syst. Sci. Data* **2016**, *8*, 605.

Li, Z.; Fox, J. M. Mapping Rubber Tree Growth in Mainland Southeast Asia Using Time-Series MODIS 250 m NDVI and Statistical Data. *Appl. Geogr.* **2012**, *32*, 420–432.

Lorenz, K.; Lal, R. Soil Organic Carbon Sequestration in Agroforestry Systems: A Review. *Agron. Sustain. Dev.* **2014**, *34*, 443–454.

Luo, Y.; White, L. W.; Canadell, J. G.; DeLucia, E. H.; Ellsworth, D. S.; Finzi, A.; Lichter, J.; Schlesinger, W. H. Sustainability of Terrestrial Carbon Sequestration: A Case Study in Duke Forest With Inversion Approach. *Glob. Biogeochem. Cycles.* **2003**, *17* (1), 1021.

Mandal, B.; Majumder, B.; Adhya, T. K.; Bandyopadhyay, P. K.; Gangopadhyay, A.; Sarkar, D.; Kundu, M. C.; Choudhury, S. G.; Hazra, G. C.; Kundu, S.; Samantaray, R. N. Potential of Double-Cropped Rice Ecology to Conserve Organic Carbon Under Subtropical Climate. *Glob. Chang. Biol.* **2008**, *14*, 2139–2151.

Manivong, V.; Cramb, R. A. Economics of Smallholder Rubber Production in Northern Laos. *Agrofor. Syst.* **2008**, *74*, 113–125.

Ngo, K. M. Carbon Stocks in Primary and Secondary Tropical Forests in Singapore. *For. Ecol. Manag.* **2013**, *296*, 81–89.

Paul, E. A.; Collins, H. P.; Leavitt, S. W. Dynamics of Resistant Soil Carbon of Midwestern Agricultural Soils Measured by Naturally Occurring 14C Abundance. *Geoderma* **2001**, *104*, 239–256.

Phuc, T. X.; Nghi, T. H. *Rubber Expansion and Forest Protection in Vietnam*; Tropenbos International Viet Nam: Hue, Viet Nam, 2014.

Plattner, G.; Joos, F.; Stocker, T. F. Revision of the Global Carbon Budget Due to Changing Air-Sea Oxygen Fluxes. *Glob. Biogeochem. Cycles.* **2002**, *16*, 1096.

Polasky, S.; Carpenter, S. R.; Folke, C.; Keeler, B. Decision-making Under Great Uncertainty: Environmental Management in an Era of Global Change. *Trends Ecol. Evolut.* **2011**, *26*, 398–404.

Prachaya, J. *Rubber Economist Quarterly Report*; The Rubber Economist Ltd.: London and Bangkok, 2009.

Prentice, I. C.; Farquhar, G. D.; Fasham, M. J. R.; Goulden, M. L.; Heimann, M.; Jaramillo, V. J.; Kheshgi, H. S.; LeQuéré, C.; Scholes, R. J.; Wallace, Douglas W. R.; The Carbon Cycle and Atmospheric Carbon Dioxide. In *Climate Change 2001: The Scientific Basis;* Houghton, J. T., Ding, Y., Griggs, D. J., Noguer, M., van der Linden, P. J., Dai, X., Maskell, K., Johnson, C. A., Eds.; Cambridge University Press: Cambridge. 2001; pp 183–237.

Roshetko, J. M.; Delaney, M.; Hairiah, K.; Purnomosidhi, P. Carbon Stocks in Indonesian Homegarden Systems: Can Smallholder Systems be Targeted for Increased Carbon Storage? *Am. J. Alternat. Agri.* **2002**, *7*, 1–11.

Rubber Statistical News, 2012. Rubber Board. Ministry of Commerce and Industry, Government of India. 2012. http://rubberboard.org.in/monstatsdisplay.asp (accessed Jan 26, 2018).

Sanchez, P. A. Linking Climate Change Research with Food Security and Poverty Reduction in the Tropics. *Agri. Ecosyst. Environ.* **2000**, *82*, 371–383.

Sharrow, S. H.; Ismail, S. Carbon and Nitrogen Storage in Agroforests, Tree Plantations, and Pastures in Western Oregon, USA. *Agrofor. Syst.* **2004**, *60*, 123–130.

Shepashenko, D.; Shvidenko, A.; Nilsson, S. Phytomass (live biomass) and Carbon of Siberian Forests. *Biomass Bioenergy.* **1998**, *14*, 21–31.

Ter-Mikaelian, M. T.; Korzukhin, M. D. Biomass Equation for Sixty-Five North American Tree Species. *For. Ecol. Manag.* **1997**, *97*, 1–24.

The Rubber Board. *Rubber Growers Companion.* Government of India, Kottayam, Kerala, India, 115, 2005.

UNFCCC. *Kyoto Protocol to the United Nations Framework Convention on Climate Change.* Adapted at COP3 in Kyoto, Japan, on December 11, 1997.

UNFCCC. Report of the Conference of the Parties on its Thirteenth session, held in Bali from 3 to 15 December 2007, Addendum Part Two: Action Taken by the Conference of the Parties at its Thirteenth Session, 2008

UNFCCC. Report of the Conference of the Parties on Its Sixteenth Session, Cancun, November 29–December 10, 2010, Addendum Part Two: Action Taken by the Conference of the Parties at Its Sixteenth Session; 2011.

Wei, X. R.; Shao, M. G.; Gale, W.; Li, L. H. Global Pattern of Soil Carbon Losses due to the Conversion of Forests to Agricultural Land. *Sci. Rep.* **2014**, *4*, 4062.

Zhai, D. L.; Cannon, C. H.; Slik, J. W. F.; Zhang, C. P.; Dai, Z. C. Rubber and Pulp Plantations Represent a Double Threat to Hainan's Natural Tropical Forests. *J. Environ. Manag.* **2012**, *96*, 64–73.

Ziegler, A. D.; Fox, J. M.; Xu, J. The Rubber Juggernaut. *Science* **2009**, *324*, 1024–1025.

CHAPTER 2

Biomass Models and Biomass Carbon Stocks

2.1 INTRODUCTION

The findings of the Intergovernmental Panel on Climate Change on global warming are more certain and alarming than ever. Since 1750, the concentration of greenhouse gases in the atmosphere has increased substantially mainly due to anthropogenic activities (Lal, 2008). Cumulative anthropogenic CO_2 (carbon dioxide) emission during 1750–2011 in the atmosphere was 555 Gt C (1 gigaton = 1 billion tons), of which 375 Gt C was emitted from fossil fuel burning and cement industries and another 180 Gt C from the deforestation and other land use changes (IPCC, 2013). Annual average CO_2 emissions over 2002–2011 from fossil fuel burning, cement industries, and anthropogenic land use changes were 8.3, 9.5, and 0.9 Gt C year^{-1}, respectively (IPCC, 2013). Annually, 11.6 million ha of natural forests are lost in the tropics due to deforestation, degradation, and desertification, especially due to increased pressure on forests from demographic, economic and social changes (FAO, 2006; Grace, 2004). More specifically in case of Southeast Asia, Houghton and Hackler (1999) showed that total emissions of carbon to the atmosphere due to land-use change and forestry changes from 1850 to 1995 were (however, at an uncertainty of 50%) 43.5 petagram (Pg) of CO_2, whereof 33.5 Pg CO_2 originated from clearing of forests into permanent croplands and 11.5 Pg CO_2 from shifting cultivation, logging, and fuel-wood extraction. However, since 1980, a steady decline of carbon emissions from land-use change in Southeast-Asia was registered by Houghton and Hackler (1999). However, the observed increase in

---

Note: This chapter has been reprinted and modified with permission from Brahma, B., Sileshi, G.W., Nath, A. J., and Das, A. K. Development and evaluation of robust tree biomass equations for rubber tree (Hevea brasiliensis) plantations in India, Forest Ecosystems (2017) 4:14. Creative Commons Attribution 4.0 International License (http://creativecommons.org/licenses/by/4.0/).

global average temperatures is "very likely" due to the steady increase of CO_2 in our atmosphere, that is, from 280 parts per million (ppm) in 1850 up to 400 ppm in 2015 (Le Quere et al., 2015; IPCC, 2007). To avoid and combat this dangerous climate change 187 nations have so far ratified the Kyoto Protocol to work on how to combat climate change in post Kyoto era (UNFCCC, 1997). In order to stabilize the carbon dioxide concentrations below 450 ppm till 2050, a numerous conventions and mechanisms have been developed so far (UNFCCC, 1997). Clean development mechanism (CDM) is one of the three mechanisms under flexible market mechanisms adopted in the KP. This mechanism encourages the role of biomass in carbon stock management through afforestation and reforestation activities. Since forestry and agroforestry systems possesses higher C stock and sequestration potential than that of pastures and field crops (Kirby and Potvin, 2007; Sharrow and Ismail, 2004; Roshetko et al., 2002; Sanchez, 2000), the former have attracted the interests of policy makers in developing policies related to removal of atmospheric CO_2 concentration through vegetation (Lorenz and Lal, 2014). Worldwide a net uptake of 1.1–1.6 Pg (1 Pg = 10^{12} g) CO_2 per year could be achieved by developing plantation forestry or agroforestry as reforestation and afforestation activities between 1995 and 2050 (IPCC, 2003). Therefore, plantation development activities could be a meaningful strategy to mitigate the atmospheric CO_2 concentration level. Since, plantation forestry systems (like rubber tree based systems) are receiving more research interest for their high carbon sink capacity (Lorenz and Lal, 2014) and according to the Food and Agriculture Organization (FAO) definition rubber tree plantations correspond to the definition of forest (World Rainforest Movement, 2010), CDM offers the possibility of enhancing rubber plantation development and socio-economic upliftment of the participating host country. Rubber is a "hot commodity" with consumption increasing worldwide at an average rate of 5.8%per year since 1900 (The Rubber Board, 2005) has boosted the area under rubber tree plantations to >1,000,000 ha in upland areas of China, Laos, Thailand, Vietnam, Cambodia, and Myanmar (Manivong and Cramb, 2008, Ziegler et al., 2009, Li and Fox, 2012). Therefore, it is important to study the role of rubber tree plantations in global atmospheric carbon concentration stabilization.

Estimation of total biomass stock involves the precise accounting of above ground and below ground biomass of the tree (IPCC, 2006). Destructive and non-destructive methods of biomass estimation are useful to obtain the approximate biomass values of trees (Brown, 1997). The destructive method includes harvesting of the whole tree followed by summing up the dry weight of different tree component to yield the dry mass value. Harvesting the whole tree is a reliable and useful to accurate estimations of tree biomass till date. However, this approach has several adverse impacts on tree resources. Therefore, development of biomass equations through destructive method and use it for future estimation is getting interest of the current biomass estimation domain study (Kumar and Nair, 2011; Chave et al., 2005; Brown, 1997). Since, for biomass estimation the first stone of biomass model development was laid down since long ago, till date a number of studies have developed different models or equations for tropical and sub-tropical trees for future biomass estimation (Chave et al., 2015, 2005; Brown, 1997; Brown and Lugo, 1992, 1984). However, major issue with many of the model is their specific use of certain independent variables like tree height and diameter for tree biomass estimation. Nonetheless, species specific model development with maximum reliability and accuracy is still need to be reported. Furthermore, due to the vegetative complexity in the associations of tropical and subtropical forests and agroforestry systems, existed allometric equations are not been so appropriate for estimating the biomass (Chave et al., 2005; Dey et al., 1996; Whittaker and Niering, 1975). Moreover, while studying with rubber, it has been suggested that the generalized equations have limitations when different agro-climatic conditions are taken into consideration (Dey et al., 1996). Therefore, it is more appropriate to develop a rubber tree biomass equation to facilitate non-destructive approach of rubber tree biomass estimation regionally and age specifically.

Rubber tree plantations are managed throughout the tropical worlds due to their significant economic value (Fox et al., 2012) and each plantation is managed under homogeneous age series up to maximum of 40 years (Brahma et al., 2016). The need for developing models covering a wide range of tree ages is important because the shape of the relationships between tree variables change along the stage of development. The variations in biomass stocks in rubber plantations with plantation age have been demonstrated (Tang et al., 2009). All these raise concerns about the applicability of available models for future biomass estimation in India. Therefore, the objective of this study was to develop biomass estimation models for different tree components in rubber plantations and assess model predictive performance at the stand level.

2.2 METHODS FOR BIOMASS DATA GENERATION

Data for biomass model development was gathered from northeast India. The main criterion for selecting the specific plantation was the age of the stand. Rubber plantations of four different ages, that is, 6, 15, 27, and 34 years, were selected for this study on the basis of the availability of the stands under the same age. Ten plots measuring 25 m × 25 m area were selected randomly from each of the different aged plantations and all trees were counted for estimation of tree density. All the trees within the selected area were counted and measured at circumference at 200 cm (C200) above the ground to avoid the tapping artifacts. When collecting the latex from rubber trees, tapping on the stems is commonly done between stem heights of 150 and 180 cm. Regular tapping resulted in significant deformations on the stem. Therefore, instead of following the recommended height to measure the stem circumference, here the stem height at 200 cm was considered. We acknowledge that this can lead to a systematic underestimation of diameter at breast height (DBH). Standard procedures were followed during the circumference measurement of each tree (Husch et al., 2003). The measured C200 values across the ages ranged between 10 and 110 cm, which were further divided into 10 cm classes. Five to six C200 classes were considered for each stand. Based on the availability of trees in these C200 classes, a total of 67 trees of different ages (15, 19, 15, and 18 for 6, 15, 27, and 34 years, respectively) were felled. Total height (H) of each harvested tree was measured to use as a covariate in biomass model development. Here, the main support to the branches from ground level to the top of the felled tree was considered as stem. In order to avoid the ambiguity of branch and stem at the fork, the higher circumference was considered as the stem. Different tree parts including foliage, branch and stems were separated and fresh weight (fw) in kg was measured in the field. For belowground biomass, coarse roots (>2.5 mm diameter) were extracted from a 1 m radius with 1 m depth around the felled tree stem. Fine root biomass could not be estimated due to difficulties in separation of fine roots from the soil. Fresh weights of extracted coarse roots were weighed after carefully removing the soil. Sub-samples of each tree part were taken to the laboratory and subsamples were oven dried at 70°C for a minimum of 72 h or until the constant dry weight was achieved. Fresh weight and dry weight (dw) ratios ($R_{dw/fw}$) for each plant part were calculated to get their respective total part dry weight (PDW) as PDW = ($R_{dw/fw}$

× fresh weight of the plant part). Total dry weight or total biomass of the tree was estimated by adding all the calculated PDW of that tree.

2.3 BIOMASS MODEL DEVELOPMENT

Use of different independent variables for developing biomass models is still a subject of empirical debate (Sileshi, 2014). Diameter at breast height (D) and total tree height (H) have been used widely for modeling scaling relationships between tree biomass components, H and DBH assuming different physical and biological first principles (Sileshi, 2014). In this study, first we developed the relationship between D200 (here after D) and H as well as biomass components with D, H, and stand density (StD). First, we explored the appropriateness of a simple power-law relationship between AGB, D, and H. This is because power-law scaling is supported by emergent theories of macroecology and thus recommended in biomass estimation to reduce the ambiguity about allometric relationships (Sileshi, 2015, 2014). Initially, we also explored the use of stand age as a covariate. However, we found stand age to be inappropriate due statistical complications such as collinearity that arise when combined with D and H. In addition, we explored stand-level variables especially StD as a biologically meaningful and potentially more predictive variable than stand age. Finally, we compared the performance of the power-law model with three other models involving D alone, H alone and their combination with StD. When fitting the models, log transformed biomass data were linearly regressed against the log transformed values of D, H, $D^2 \times H$ and $D \times H \times StD$ as follows:

$\ln(Y) = \ln(\alpha) + \beta(\ln D) + \varepsilon$ Model 1
$\ln(Y) = \ln(\alpha) + \beta(\ln D^2 \times H) + \varepsilon$ Model 2
$\ln(AGB) = \ln(\alpha) + \beta(\ln H) + \varepsilon$ Model 3
$\ln(Y) = \ln(\alpha) + \beta(\ln D \times H \times StD) + \varepsilon$ Model 4

In addition, we explored the relationship between coarse root dry weight (RDW) and AGB using the power-law model specified as follows:

$\ln(RDW) = \ln(\alpha) + \beta(\ln AGB) + \varepsilon$ Model 5

where Y is the biomass component (e.g., foliage, branch, stem, root, etc.) being modeled, $D^2 \times H$ is tree volume, and ε is the error. These models were fitted using ordinary least square regression assuming a power function with multiplicative error structure (Dong et al., 2016; Lai et al., 2013). For estimating biomass components in the arithmetic domain, the equations were back-transformed to give $Y = \alpha X^\beta \times CF$ where X is the single or compound variables, $\alpha X^\beta = \exp(\alpha + \beta(\ln X))$ and CF is the correction factor calculated from the mean square of error (MSE) as $\exp(MSE/2)$ (Sileshi, 2015).

In the absence of measured tree height, the H in model 2 may be estimated using height–diameter (H–D) models. This is a convenient approach because data on tree height are relatively more difficult and time consuming to obtain than D. H–D models can also be used to predict missing heights from field measurement of tree diameter. The commonly used H–D models include Chapman–Richard, exponential, Gompertz, hyperbolic, logistic, Michaelis–Menten (saturation growth), monomolecular, power law, Richard and Weibull functions (Huang et al., 2000; Zeide, 1993). The models are also applied to estimate the asymptotic height, which is often used as a measure of tree size (Thomas, 1996). In order to determine the appropriate H–D relationship in rubber trees we compared the following models:

Chapman–Richards $\quad H = a\left(1 - \exp(-bD)\right)^c$

Exponential $\quad H = a - b\left(\exp(-cD)\right)$

Gompertz $\quad H = a\left(\exp\left(-\exp(b - cD)\right)\right)$

Hyperbolic models $\quad H = a + \dfrac{b}{D}$

Logistic $\quad H = \dfrac{a}{\left(1 + \exp(b - cD)\right)}$

Michaelis–Menten $\quad H = \dfrac{aD}{b + D}$

Monomolecular $\quad H = a\left(1 - \exp(-b(D - c))\right)$

Power-law $\quad H = aD^b$

Richards function $\quad H = \dfrac{a}{\left(1 + \exp(b - cD)\right)^{1/d}}$

Weibull $\quad H = a - b\left(\exp\left(-cD^d\right)\right)$

These models were chosen because they are supported by theory and their parameters have biologically meaningful interpretations. For example, in all models except the power-law, the parameter *a* represents the asymptotic height of trees. Nonlinear regression was used to estimate parameters of all H–D models.

The performance of all these models was compared using the coefficient of determination (R^2), Akaike information criterion (AIC), and root mean square of error (RMSE) (Sileshi, 2014). In all cases, the small sample approximation of AIC (i.e., AICc) was used for comparing models. In addition, we checked the normality of residuals by plotting the residuals against the predictor values and the Shapiro–Wilk test. Finally, we compared the 95% confidence intervals (CI) of the slopes to see whether the corresponding model for different age classes is significantly different or not (Sileshi, 2015, 2014).

Using the best model developed through the steps described above, biomass stocks in plantations of different ages was estimated by using the following formula:

Biomass density (Mg ha^{-1}) = $N \times (B_1 + B_2 + B_3 + \ldots\ldots\ldots + B_n)/n$,

where *N* is the tree density, *n* is the number of harvested trees, and B_i is the AGB stock (kg) of the respective tree.

2.4 BIOMASS ESTIMATION MODELS

Tree density ranged from 784 trees ha^{-1} in 6-year-old plantations to 576 trees ha^{-1} in 34-year-old plantations (Table 2.1). Stem diameter and tree height also significantly ($P < 0.0001$) varied with stand age. Tree diameter ranged from 11 to 35 cm in 6 and 34-year-old plantations, respectively (Table 2.1).

TABLE 2.1 Average Tree Density, Stem Diameter at 200 cm, Height (cm) and Aboveground Biomass (AGB), Coarse Root Biomass and Total Biomass Stocks Estimated Using the Best Model (Model 2) in Different Ages of Rubber Plantation in Northeast India.

Stand age (years)	Density (tree ha^{-1})	Diameter (cm)	Height (cm)	AGB (Mg ha^{-1})	Coarse root (Mg ha^{-1})	Total biomass (Mg ha^{-1})
6	784 (6)	11.1 (1.48)	8.3 (0.68)	28.9 (18.9)	4.1 (1.0)	31.5 (18.2)
15	720 (11)	19.9 (1.32)	15.2 (0.61)	119.4 (16.8)	10.3 (0.9)	121.4 (16.2)
27	688 (11)	19.5 (1.48)	16.5 (0.68)	116.4 (18.9)	10.0 (1.0)	118.4 (18.2)
34	576 (25)	23.8 (1.35)	20.0 (0.62)	169.6 (17.3)	11.9 (1.0)	167.6 (16.6)
All ages	692 (44)	18.9 (0.88)	15.3 (0.60)	112.0 (10.7)	9.3 (0.6)	113.0 (10.3)

Values in parentheses are standard errors of means.

TABLE 2.2 Parameter Estimates and Model Fit Criteria for 10 Different Height–Diameter (H–D) Models.

Model	a	b	c	d	AICc	R^2	RMSE	95% CI of asymptotic height
Michaelis–Menten	51.1	42.3	–[a]	–	**112.4**	**0.795**	**2.24**	33.8–68.5
Power law	1.98	0.70	–	–	114.3	0.790	2.27	–
Gompertz	24.9	0.84	0.09	–	114.4	0.796	2.25	20.0–29.8
Chapman–Richard	30.0	0.04	1.07	–	114.5	0.796	2.25	13.5–46.5
Monomolecular	31.3	0.04	0.17	–	114.5	0.796	2.25	18.2–44.4
Exponential	31.3	31.52	0.04	–	114.5	0.796	2.25	18.2–44.4
Logistic	23.1	1.72	0.13	–	114.9	0.795	2.26	19.8–26.3
Weibull	25.5	23.40	0.02	1.34	116.6	0.797	2.27	11.2–39.9
Richard[b]	24.8	−2.12	0.09	0.05	116.8	0.796	2.27	14.0–35.5
Hyperbolic	20.9	−84.9	–	–	159.7	0.585	3.19	19.5–22.3

[a]Indicates that parameter is not applicable.
[b]Indicates that convergence criteria not met.

A strong non-linear relationship was observed between stem height (H) and diameter at 200 cm (D) in all the stand ages (Fig. 2.1; Table 2.2).

Among the H–D models compared, the Michaelis–Menten function had the smallest AIC and RMSE and the largest pseudo R^2. The second best model was the power-law model, while the worst was the hyperbolic function. The differences in terms of model goodness of fit criteria (pseudo R^2, AIC, and RMSE) among the models except the hyperbolic were marginal (Table 2.3).

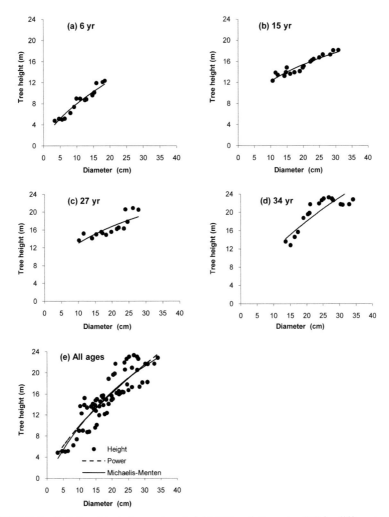

FIGURE 2.1 Relationship between tree height (H) and diameter (D) in different aged stands (a–d) and generalized across plantation ages (e). Parameters of the fitted lines (power-law) are presented in Table 2.2.

TABLE 2.3 Parameters[a] and Model Fit Statistics for the Age-Specific Models of Aboveground Biomass (AGB). For Ease of Comparing Slopes (β) the 95% Confidence Limits Were Presented.

Model	Stand age (year)	Intercept (ln(α))	Slope (β)[b]	Model R^2	Fit AIC	Criteria RMSE
1	6	−2.9	2.57 (2.49–2.65)	**0.997**	**−70.8**	**0.077**
	15	−2.01	2.25 (2.11–2.39)	**0.984**	**−84.0**	**0.095**
	27	−0.32	1.84 (1.53–2.15)	0.909	−45.5	0.717
	34	−2.12	2.43 (2.18–2.68)	0.958	−64.3	0.143
	All ages	−2.83	2.60 (2.46–2.74)	0.954	−170.6	0.271
2	6	−3.49	0.97 (0.94–1.00)	0.998	−66.6	0.089
	15	−3.64	0.96 (0.90–1.02)	0.980	−79.4	0.107
	27	−1.71	0.78 (0.67–0.89)	**0.938**	**−51.4**	**0.059**
	34	−3.08	0.93 (0.87–0.99)	**0.983**	**−80.5**	**0.092**
	All ages	−3.31	0.95 (0.91–0.99)	0.976	−217.1	0.192
3	6	−4.85	3.79 (3.22–4.36)	0.93	−24.2	0.365
	15	−11.28	5.85 (4.71–6.99)	0.858	−42.3	0.285
	27	−5.53	3.79 (2.73–4.85)	0.79	−32.8	0.274
	34	−4.15	3.23 (2.63–3.83)	0.876	−44.8	0.246
	All ages	−3.49	3.04 (2.81–3.27)	0.911	−126.6	0.376

[a] All parameters were significantly different from zero. Therefore, P values were not included.
[b] Figures in parentheses represent 95% confidence limits of the slope. Two or more slopes are significantly different from each other only if their confidence limits do not overlap.
Model fit statistics of the best model for a given plantation age are in boldface.

However, the 95% confidence limits of the estimated asymptotic heights reveal significant differences among the models. Some of the models (e.g., Weibull and Richard functions) had four parameters and hence are complicated or fail to converge. For practical purposes, the following two models were chosen for estimating H in the absence of measured height:

Michaelis–Menten: $H = \dfrac{51.1D}{42.3 + D}$

Power–law: $H = 1.99 D^{0.70}$

Examination of the confidence limit of the slopes of the age-specific models (Table 2.3) indicated that slopes for ages 6 and 34 years are not significantly different from the slopes of the models generalized across all age classes. Due to this and the small sample size for individual ages, equations that are generalized across all age classes were developed and tested. Based on the higher R^2 and lower AIC and RMSE, Models 2 and 4 were found to be more appropriate for AGB, stem, coarse root and total dry weight than Models 1 and 3 (Table 2.4).

TABLE 2.4 Parameters and Goodness of Fit Statistics for the Various Models used in Estimating Biomass Components Combining all Stand Ages.

Components	Model	Intercept	Slope	R^2	AICc	RMSE
AGB	1	−2.82 (0.21)*	2.60 (0.07)	0.954	−170.6	0.270
	2	−3.31 (0.15)	0.95 (0.02)	**0.977**	**−217.1**	**0.192**
	3	−3.49 (0.32)	3.03 (0.12)	0.911	−126.6	0.377
	4	−14.47 (0.46)	1.58 (0.04)	0.963	−186.2	0.241
Foliage	1	−4.41 (0.51)	1.87 (0.18)	**0.635**	**−49.0**	**0.672**
	2	−4.48 (0.57)	0.65 (0.07)	0.588	−40.9	0.713
	3	−3.91 (0.71)	1.82 (0.27)	0.418	−17.8	0.847
	4	−11.89 (1.41)	1.07 (0.12)	0.559	−36.4	0.738
Branches	1	−2.76 (0.46)	2.03 (0.16)	**0.714**	**−62.5**	**0.607**
	2	−2.95 (0.51)	0.71 (0.06)	0.687	−56.4	0.636
	3	−2.60 (0.65)	2.11 (0.24)	0.543	−31.1	0.767
	4	−11.32 (1.25)	1.19 (0.10)	0.672	−53.4	0.650
Stem	1	−3.97 (0.27)	2.88 (0.09)	0.934	−131.8	0.362
	2	−4.57 (0.19)	1.05 (0.02)	**0.972**	**−188.4**	**0.237**
	3	−4.95 (0.29)	3.45 (0.11)	0.941	−139.4	0.342
	4	−17.09 (0.50)	1.77 (0.04)	0.967	−176.5	0.259
Coarse roots	1	−2.4 (0.23)	1.7 (0.08)	0.871	−155.1	0.305
	2	−2.64 (0.23)	0.60 (0.03)	0.883	−161.9	0.290
	3	−2.69 (0.32)	1.91 (0.12)	0.802	−126.7	0.375
	4	−9.84 (0.56)	1.02 (0.05)	0.881	−160.9	0.292
	5	−0.53 (0.12)	0.64 (0.03)	**0.905**	**−175.4**	**0.261**
Total biomass	1	−2.38 (0.20)	2.49 (0.07)	0.954	−176.7	0.259
	2	−2.84 (0.15)	0.90 (0.02)	**0.976**	**−221.2**	**0.184**
	3	−3.00 (0.31)	2.89 (0.11)	0.908	−130.9	0.365
	4	−13.50 (0.45)	1.51 (0.04)	0.963	−191.1	0.232

*Figures in parentheses are standard errors of estimates; Goodness of fit statistics of the best model is in boldface.

On the other hand, the power-law model was more appropriate for foliage and branch biomass. Therefore, the following models were proposed for estimation of AGB, foliage, branch, stem, coarse root (RDW), and total biomass components:

$AGB = (\exp(-3.31 + 0.95(\ln D^2 \times H))) \times 1.02$

$Foliage = (\exp(-4.41 + 1.87(\ln D))) \times 1.25$

$Branch = (\exp(-2.76 + 2.07(\ln D))) \times 1.20$

$Stem = (\exp(-4.57 + 1.05(\ln D^2 \times H))) \times 1.03$

$RDW = (\exp(-0.53 + 0.64(\ln AGB))) \times 1.03$

$Total = (\exp(-2.84 + 0.90(\ln D^2 \times H))) \times 1.02$

In addition, RDW could be estimated using Model 2 as $RDW = (\exp(-2.64 + 0.60(\ln D^2 \times H))) \times 1.04$. However, in terms of goodness of fit criteria, Model 5 was only slightly better than Model 2 and Model 4 (Table 2.4).

2.5 VARIATION IN TREE BIOMASS COMPONENTS

The measured and estimated biomass components significantly varied with stand age (Fig. 2.2). However, the measured biomass components were not significantly different from the estimated biomass in all plantation ages. The exception was branch dry weight at 34 years, where the estimated branch dry weight was higher than the measured dry weight. Measured total aboveground (AGB) dry weight increased from 35.4 kg tree^{-1} in 6 year plantations to 288.5 kg tree^{-1} in 34 year old plantations (Fig. 2.2a). However, using different models the estimated total AGB for all the ages are given in Figure 2.3.

Biomass Models and Biomass Carbon Stocks

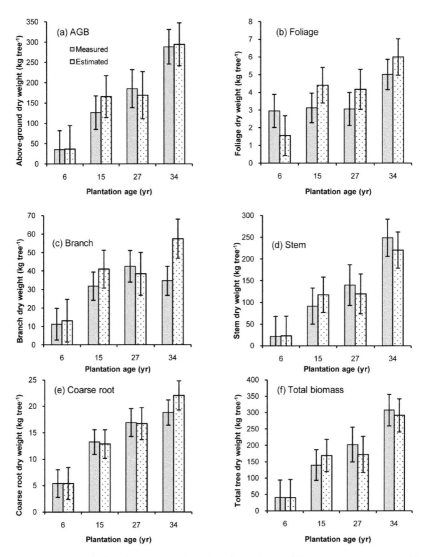

FIGURE 2.2 Variation in measured and estimated tree biomass components with plantation age. Parameters of the best model in Table 2.4 were used for estimating the specific biomass component. The measured and estimated biomass components are deemed significantly different from each other only if their confidence limits do not overlap.

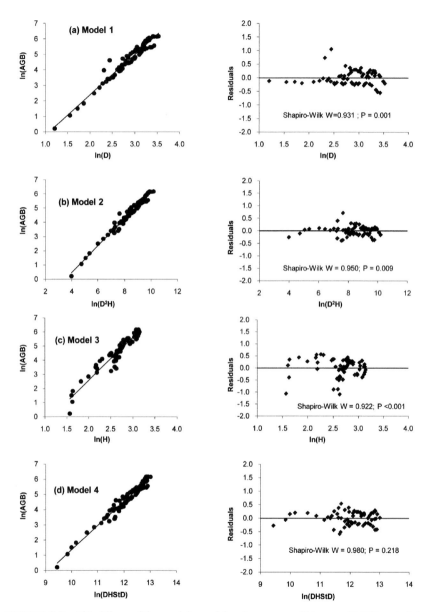

FIGURE 2.3 Fitted lines of the models used for above-ground biomass (AGB) and plots of the residuals against observed values for rubber plantations of all ages. Parameters of the fitted lines are presented in Table 2.4.

Measured foliage dry weight increased from 3 kg tree^{-1} in 6-year-old plantations to 5 kg tree^{-1} in 34-year-old plantations, but foliage biomass in 6 years plantations did not significantly differ from those in 27 years plantations (Fig. 2.2b). Measured branch dry weight increased from 11 kg tree^{-1} in 6 years plantations to 43 kg tree^{-1} in 27-year-old plantations and then decreased to 35 kg tree^{-1} in 34-year-old plantations (Fig. 2.2c). Measured stem dry weight increased from 21 kg tree^{-1} in 6 years plantations to 249 kg tree^{-1} in 34-year-old plantations (Fig. 2.2d). Similarly, coarse root dry weight increased from 5.4 kg tree^{-1} in 6 years plantations to 19 kg tree^{-1} in 34-year-old plantations (Fig. 2.2e). The coarse root biomass estimated using the three models also did not significantly differ (Fig. 2.4). However, estimating coarse root biomass from AGB (Model 5) was superior to other models in terms of goodness of fit criteria (Table 2.4) and normality of residuals (Fig. 2.4c).

Total dry weight of all components increased from 40 kg tree^{-1} in 6-year-old plantations to 307 kg tree^{-1} in 34-year-old plantations (Fig. 2.2e). In terms of the mean absolute percentage error (MAPE), AGB, total biomass and stem biomass were estimated with less error (MAPE < 20%) compared to coarse root dry weight (MAPE = 21%). Branch dry weight (MAPE = 80%) and foliage dry weight (MAPE = 83%) were poorly estimated. Foliage dry weight was underestimated in 6-year-old plantations compared to the measured (Fig. 2.2b), while stem dry weight was severely overestimated in 34-year-old plantations.

The estimated AGB stocks (Mg ha^{-1}) were 29, 119, 116, and 169 Mg ha^{-1} for 6, 15, 27, and 34-year-old plantations, respectively (Table 2.1). The coarse root biomass was estimated at 4 Mg ha^{-1} for 6-year-old plantations and 12 Mg ha^{-1} for 34-year-old plantations. Similarly, total biomass (excluding fine roots) was estimated at 32 Mg ha^{-1} at 6 years and 168 Mg ha^{-1} at 34-year-old plantations (Table 2.1).

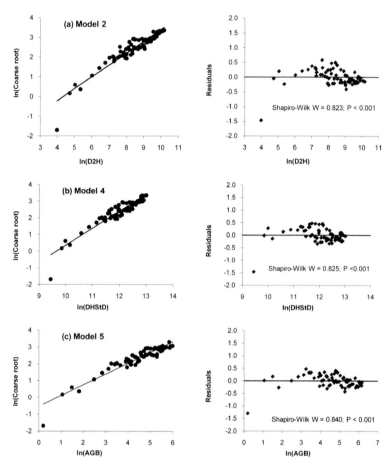

FIGURE 2.4 Fitted lines of the models used for coarse root dry weight (RDW) and plots of the residuals against observed values for rubber plantations of all ages. Parameters of the fitted lines are presented in Table 2.4.

2.6 MEASURED BIOMASS CARBON STOCK AND SEQUESTRATION RATE

Total vegetation carbon stock (above and below ground) (Mg ha^{-1}) ranged from 16.00 (6 years) to 105.73 (34 years) (Fig. 2.5). Contribution of above and below ground to total vegetation carbon stock was 13.88–99.25 Mg ha^{-1} and 2.12–6.47 Mg ha^{-1}, respectively (6–34 years) (Fig. 2.5). Analysis of vegetation carbon sequestration rate revealed annually 3.20 Mg ha^{-1}

year^{-1} organic carbon is being accumulated in rubber tree plantations under 6–34 year age (Fig. 2.6).

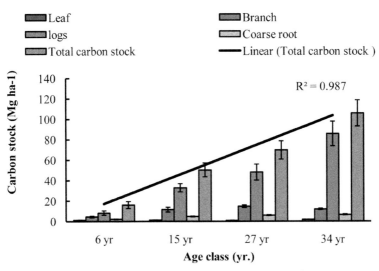

FIGURE 2.5 **(See color insert.)** Measured Carbon stock (Mg ha^{-1}) under different aged rubber tree plantations.

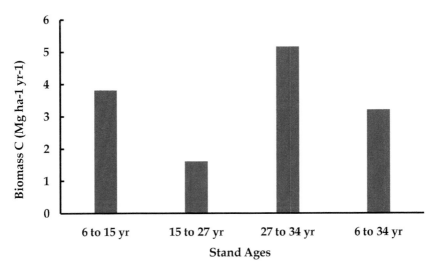

FIGURE 2.6 Biomass carbon sequestration rate under different age class of rubber tree plantations.

2.7 GLOBAL STATUS OF BIOMASS CARBON STOCK IN RUBBER PLANTATIONS

The average StD of rubber plantations declined from 784 trees ha^{-1} in 6-year-old plantations to 576 in 34-year-old plantations. Since a common plant spacing of 3 m × 4 m was maintained during the initial plantation stage, the results indicate that the number of trees in plantation decreased by about 26% with the increase in stand age. This implies that plantation density reduced at 0.95% per annum, and this is comparable with the 1.5% reduction reported from Sri Lanka (Munasinghe et al., 2014). This is consistent with predictions of Yoda's law (Yoda et al., 1963) and the StD rule of Reineke (Reineke, 1933).

From the H–D models, it is evident that the Michaelis–Menten and power-law models are adequate, while the gain in explanatory power from the other models was minimal. Therefore, in the absence of measured tree height, H estimated using the Michaelis–Menten function can be used to estimate tree volume. The results also suggest that AGB can be estimated more accurately using models consisting of compound variables of D and H. The power-law model (Model 1) was slightly poorer in terms of goodness of fit criteria. The poor performance of the power-law model could be attributed to data truncation (Sileshi, 2015), which is evident in Figure 2.1. Data truncation can occur either due to sampling that only includes those individuals whose size lies within a certain interval or when equations are fitted to a restricted segment of the size range. Data are said to be left truncated when the lower segment of the population has been left out, whereas right-truncation results from leaving out the upper segment. Sileshi (2015) has demonstrated that the allometry exponent can be highly biased when the power-law equation is fitted to a restricted segment of the size range or different life stages of a species. In the different plantation ages except at 6 years, data are evidently left truncated, that is, the lower segment of the population has been left out. Left-truncation of data tends to raise the allometry intercept and as a result shrink the slope (Sileshi, 2015). On the other hand, Model 2 slopes did not significantly differ from each other except at plantation age of 15. As the models for all ages were developed from a larger dataset, they may be less sensitive to change in age and hence may be used for future AGB estimations of 6, 27, and 34 years old plantations.

Among the models compared, the model that contains tree volume had higher predictive power for AGB and total tree biomass. This is consistent with the general observation that AGB increases with bole volume in woody species (Chave et al., 2005). Although the use of fewer explanatory variables is recommended for ease in model application and validation (Sileshi, 2014), in the present study D and H were directly measured and therefore could be applied for biomass estimation.

In this study, the focus was on the aboveground living biomass and coarse roots (excluding fine roots) of trees since these components account for the largest percentage of the sequestered carbon within a forest ecosystem (Kongsager et al., 2013). Root biomass is difficult to measure in most conditions. Therefore, we recommend direct estimation of coarse root biomass from measured AGB assuming a power-law relationship between root biomass and AGB (i.e., Model 5). This assumption is reasonable because allometry theory and empirical observations have demonstrated near-isometric relationships between aboveground and belowground biomass of trees (Cheng and Niklas, 2007; Hui et al., 2014).

The estimated AGB stocks (Mg ha^{-1}) for 6 and 15 years old rubber trees are comparable with values (56 and 248 kg tree^{-1}) reported from Brazil (Maggiotto et al., 2014). AGB estimates for 27- and 34-year-old plantations are 25% and 50% lower than the estimated total biomass for 25- and 38-year-old rubber plantations from China (Tang et al., 2009; Yang et al., 2014). These differences could be related to differences in planting density, growth habitat and management practices (Yang et al., 2014). Moreover, total biomass stocks of 5–40 years old rubber plantations from the same region were reported to be 32–211 Mg ha^{-1} (Brahma et al., 2016). Similar to the present study the AGB stock in rubber plantations increased with increase in plantation age (Brahma et al., 2016; Corpuz et al., 2014; Maggiotto et al., 2014; Tang et al., 2009; Yang et al., 2005; Dey et al., 1996; Chaudhuri et al., 1995).

The biomass carbon stock under rubber tree plantations increased with the increase of plantation age (Petsri et al., 2013). In the present study, the total biomass carbon stock of rubber plantations for 6, 15, 27, and 34 years were estimated at 16.00 (±3.38), 50.29 (±6.73), 69.59(±8.84) and 105.73(±12.71) respectively. Moreover, the findings of the present study have been supported by reports of other researchers. Biomass carbon stocks of rubber tree plantations of below 10, 10–20, 20–30 and above 30-year-old stands from different regions of the world were ranges from

12 to 22 Mg ha⁻¹, 35 to 76 Mg ha⁻¹ 66 to 78 Mg ha⁻¹ and 86 to 170 Mg ha⁻¹, respectively (Petsri et al., 2013; Wauters et al., 2008; Corpuz et al., 2014; Huber et al., 2012; Maggiotto et al., 2014; Yang et al., 2005). The carbon stock under 27 years age class of rubber plantation was higher than that of semiarid, sub-humid, humid and temperate regions agroforestry systems (Montagnini and Nair, 2004). Moreover, carbon stock for 34-year-old of rubber plantation falls within the range of tropical forests of northeast India (97–149 Mg ha⁻¹) (Upadhaya et al., 2015) and mango (*Mangifera indica*) agroforestry systems of Indonesia (121 Mg ha⁻¹) (Kirsfianti et al., 2002).

The rate of carbon sequestration in biomass of a tree mainly varies with the age and adopted management activities (Wauters et al., 2008). In the present study, the carbon sequestration rate under rubber plantations is ranged from 1.61 to 5.16 Mg C ha⁻¹ year⁻¹. However, the estimated rate of carbon sequestration in the present study was found to be lower/comparable than/to that of under 14–38 year plantations of Brazil and China 5.6–7.6 Mg ha⁻¹ year⁻¹ (Wauters et al., 2008; Huber et al., 2012; Yang et al., 2005). The estimated mean biomass carbon sequestration rate (across all the plantation ages) was 3.20 Mg ha⁻¹ year⁻¹ was found to be higher than that of under 20–25 years rubber stands of Thailand is 1.71 Mg ha⁻¹ year⁻¹ (Petsri et al., 2013). In addition to that, carbon accumulation rate observed in northeast India is comparable with the community based teak forests in Madhya Pradesh, India (3.4 Mg ha⁻¹ year⁻¹) (Poffenberger et al., 2001) but lower than many tropical agroforestry and forestry systems (Prasad et al., 2012). Although carbon sequestration rate in rubber plantation is comparable to many agroforestry and plantations systems with high economic profitability from latex production assures its long-term management and therefore, acts as a permanent sink of atmospheric CO_2.

2.8 CONCLUSIONS

It is concluded that models involving tree volume are more appropriate for regional level biomass estimation than simple models for individual stands. We recommend that the power-law model should not be used for estimation of AGB of plantations at different growth stages because power-law parameters can be biased due to data truncation. Our study suggests that the model generalizing all the ages can be used irrespective of the age of the stand. In the absence of measured height, we recommend the Michaelis–Menten model for estimating height. One of the limitations of this study is the relatively small sample size for developing age-specific models. Another limitation is our inability to sample total belowground biomass due to resource constraints to collect fine roots. We recommend that future studies tackle these issues.

From the biomass carbon stock sequestration results it is concluded that (1) biomass carbon stock in rubber plantation is comparable or even more than many tropical and subtropical forestry and agroforestry systems, (2) adopting best management practices may further enhance carbon sequestration rate in rubber tree plantations, and (3) expansion of rubber plantations can offer an ecological stability over the increasingly practiced traditional slash and burn agricultural system in Southeast Asia, concurrently uplifting the socioeconomic conditions of local people through generating income stream from carbon trading.

KEYWORDS

- **allometric models**
- **H-D models**
- **biomass density**
- **coarse root**
- **ecological stability**

REFERENCES

Brahma, B.; Nath, A. J.; Das, A. K. Managing Rubber Plantations for Advancing Climate Change Mitigation Strategy. *Curr. Sci.* **2016**, 110, 2015–2019.

Brown, S. *Estimating Biomass and Biomass Change in Tropical Forest: A Premier UN FAO Forestry Paper;* FAO: Rome, 1997 (134); p 55.

Brown, S.; Lugo, A. E. Biomass of Tropical Forests: A New Estimate Based on Forest Volumes. *Science* **1984**, *223*, 1290–1293.

Brown, S.; Lugo, A. E. Above Ground Biomass Estimates for Tropical Moist Forests of the Brazilian Amazon. *Interciencia* **1992**, *17*, 8–18.

Chaudhuri, D.; Vinod, K. K.; Potty, S. N.; Sethuraj, M. R.; Pothen, J.; Reddy, Y. A. N. Estimation of Biomass in *Hevea* Clones by Regression Method: Relation Between Girth and Biomass. *Ind. J. Nat. Rubb. Res.* **1995**, *8*, 113–116.

Chave, J.; Andalo, C.; Brown, S.; Cairns, M. A.; Chambers, J. Q.; Eamus, D.; Fölster, H.; Fromard, F.; Higuchi, N.; Kira, T.; Lescure, J. P.; Nelson, B. W.; Ogawa, H.; Puig, H.; Riéra, B.; Yamakura, T. Tree Allometry and Improved Estimation of Carbon Stocks and Balance in Tropical Forests. *Oecologia* **2005**, *145*, 87–99.

Chave, J.; Rejou-Mechain, M.; Burquez, A.; Chidumayo, E.; Colgan, M. S.; Delitti, W. B. C.; Duque, A.; et al. Improved Allometric Models to Estimate the Aboveground Biomass of Tropical Trees. *Glob. Chang. Biol.* **2015**, *20*, 3177–3190.

Cheng, D.; Niklas, K. Above- and Below-Ground Biomass Relationships Across 1543 Forested Communities. *Ann. Bot.-London.* **2007**, 99, 95–102.

Corpuz, O. S.; Abas, E. L.; Salibio, F. C.; Potential Carbon Storage of Rubber Plantations. *Ind. J. Pharm. Biol. Res.* **2014**, *2*, 73–82.

Dey, S. K.; Chaudhuri, D.; Vinod, K. K.; Pothen, J.; Sethuraj, M. R. Estimation of Biomass in *Hevea* Clones by Regression Method: 2. Relation of Girth and Biomass for Mature Trees of Clone RRIM 600. *Ind. J. Nat. Rubb. Res.* **1996**, *9*, 40–43.

Dong, L.; Zhang, L.; Li, F. Developing Two Additive Biomass Equations for Three Coniferous Plantation Species in Northeast China. *Forests*. **2016**, *7*, 136.

FAO. Global Forest Resources Assessment 2005—Progress Toward Sustainable Forest Management; FAO Forestry Paper No. 147, Rome, 2006. www.fao.org/docrep/008/a0400e/a0400e00.htm (accessed Oct 9, 2009).

Fox, J.; Vogler, J. B.; Sen, O. L.; Giambelluca, T. W.; Ziegler, A. D. Simulating Land-Cover Change in Montane Mainland Southeast Asia. *Environ. Manag.* **2012**, *49*, 968–979.

Grace, J. Understanding and Managing the Global Carbon Cycle. *J. Ecol.* **2004**, *92*, 189–202.

Houghton, R. A.; Hackler, J. L. Emissions of Carbon from Forestry and Land-Use Change in Tropical Asia. *Glob. Chang. Biol.* **1999**, *5* (4), 481–492.

Huang, S.; Price, D.; Titus, S. J. Development of Ecoregion-Based Height-Diameter Models for White Spruce in Boreal Forests. *For. Ecol. Manag.* **2000**, *129*, 125–141.

Huber, M.; Claudia, A.; Sergey, B.; Attachai, J.; Georg, C. Carbon Sequestration of Rubber (*Hevea brasiliensis*) Plantations in the Naban River Watershed National Nature Reserve in Xishuangbanna, China; Georg-August Universitat Gottingen and University of Kassel-Witzenhausen, Tropentag: Gottingen, Germany, 2012. 2012. www.tropentag.de/2012/abstracts/full/467.pd (accessed Jul 19, 2013).

Hui, D.; Wang, J.; Shen, W.; Le, X.; Ganter, P.; Ren, H. Near Isometric Biomass Partitioning in Forest Ecosystems of China. *PLoS One* **2014**, *9*, 86550.

Husch, B.; Beers, T, W.; Kershaw, J. A.; *Forest Mensuration,* 4th ed.; Wiley: Hoboken, 2003; pp 85–86.

IPCC. Good Practice Guidance for Land Use. Land-Use Change and Forestry; IPCC National Greenhouse Gas Inventories Programme, Kanagawa, Japan, 2003.

IPCC. IPCC Guidelines for National Greenhouse Gas Inventories, Prepared by the National Greenhouse Gas Inventories Programme; IGES, Japan. Eggleston, H. S., Buendia, L., Miwa, K., Ngara, T., Tanabe, K. (Eds.), 2006.

IPCC. Summary for Policymakers. In *Climate Change 2007: The Physical Basis. Contribution of Working Group I to the Fourth Assessment Report of the Intergovernmental*

Panel on Climate Change; Cambridge University Press: Cambridge, United Kingdom and New York, NY, USA, 2007.

IPCC. Intergovernmental Panel on Climate Change Summary for Policymakers. In *Climate Change 2013: The Physical Science Basis, Contribution of Working Group I to the Fifth Assessment Report of the Intergovernmental Panel on Climate Change*; Cambridge University Press: Cambridge, United Kingdom and New York, NY, USA, 2013.

Kirby, K. R.; Potvin, C. *Variation in Carbon Storage Among Tree Species: Implications for the Management of a Small-scale Carbon Sink Project. For. Ecol. Manag.* **2007**, *246*, 208–221.

Kirsfianti, G.; Yuliana, C. W.; Mega, L. Potential of Agroforestry and Plantation Systems in Indonesia for Carbon Stocks: An Economic Perspective. Working Paper CC14, ACIAR Project ASEM, 2002.

Kongsager, R, Napier, J.; Mertz, O. The Carbon Sequestration Potential of Tree Crop Plantations. *Mitig. Adapt. Strateg. Glob. Chang.* **2013**, *18*, 1197–1213.

Kumar, B. M.; Nair, P. K. R. *Carbon Sequestration Potential of Agroforestry 3 Systems: Opportunities and Challenges*. In *Advances Agroforestry;* Springer: The Netherlands, 2011; Vol. 8, p 307.

Lai, X.; Li, F.; Leung, K . A Monte Carlo Study of the Effects of Common Method Variance on Significance Testing and Parameter Bias in Hierarchical Linear Modeling. *Organizat. Res. Method.* **2013**, *16*, 243–269.

Lal, R. Sequestration of Atmospheric CO_2 in Global Carbon Pools. *Energ. Environ. Sci.* **2008**, *1*, 86–100.

Le Quéré, C.; Moriarty, R.; Andrew, R. M.; Canadell, J. G.; Sitch, S.; Korsbakken, J. I.; et al. Global Carbon Budget 2015, Earth System Science Data, 2015, *7*, 396.

Li, Z.; Fox, J. M. Mapping Rubber Tree Growth in Mainland Southeast Asia Using Time-Series MODIS 250 m NDVI and Statistical Data. *Appl. Geol.* **2012**, *32*, 420–432.

Lorenz, K.; Lal, R. Soil Organic Carbon Sequestration in Agroforestry Systems. A Review. *Agron. Sust. Dev.* **2014**, *34*, 443–454.

Maggiotto, S. R., Oliveira, D. D.; Marur, C. J.; Stivari, S. M. S.; Leclerc, M.; Wagner-Riddle, C. Potential Carbon Sequestration in Rubber Tree Plantations in the Northwestern Region of the Paraná State, Brazil. *Acta Scientiarum. Agron.* **2014**, *36*, 239–245.

Manivong, V.; Cramb,R. A. Economics of Smallholder Rubber Production in Northern Laos. *Agrofor. Syst.* **2008**, *74*, 113–125.

Montagnini, F.; Nair, P. K. R. Carbon Sequestration: An Underexploited Environmental Benefit of Agroforestry Systems. *Agrofor. Syst.* **2004**, *61*, 281–295.

Munasinghe, E. S.; Rodrigo, V. H. L.; Gunawardena, U. A. D. P. Modus Operandi in Assessing Biomass and Carbon in Rubber Plantations Under Varying Climatic Conditions. *Exp. Agric.* **2014**, *50*, 40–58.

Petsri, S.; Chidthaisong, A.; Pumijumnong, N.; Wachrinrat, C. Greenhouse Gas Emissions and Carbon Stock Changes in Rubber Tree Plantations in Thailand from 1990 to 2004. *J. Clean. Prod.* **2013**, *52*, 61–70.

Poffenberger, M.; Ravindranath, N. H.; Pandey, D. N.; Murthy, I. K.; Bist, R.; Jain, D. Communities and climate change: the clean development mechanism and village based restoration in Central India: a case study from Harda forest Division, Madhya Pradesh, India. *Community Forestry International and Indian Institute of Forest Management*, Santa Barbara. 2001.

Reineke, L. H. Perfecting a Stand-density Index for Even-aged Forests. *J. Agri. Res.* **1933**, *46*, 627–638.

Roshetko, J. M.; Delaney, M.; Hairiah, K.; Purnomosidhi, P. Carbon Stocks in Indonesian Homegarden Systems: Can Smallholder Systems be Targeted for Increased Carbon Storage? *Am. J. Altern. Agri.* **2002**, *7*, 1–11.

Sanchez, P. A. Linking Climate Change Research with Food Security and Poverty Reduction in the Tropics. *Agri. Ecosys. Environ.* **2000**, *82*, 371–383.

Sharrow, S. H.; Ismail, S. Carbon and Nitrogen Storage in Agroforests, Tree Plantations, and Pastures in Western Oregon, USA. *Agrofor. Syst.* **2004**, *60*, 123–130.

Sileshi, G. W. A Critical Review of Forest Biomass Estimation Models, Common Mistakes and Corrective Measures. *For. Ecol. Manag.* **2014**, *329*, 237–254.

Sileshi, G. W. The Fallacy of Retification and Misinterpretation of the Allometry Exponent. **2015**. doi: 10.13140/RG.2.1.2636.9768.

Tang, J. W.; Pang, J. P.; Chen, M. Y.; Guo, X. M.; Zeng, R. Biomass and Its Estimation Model of Rubber Plantations in Xishuangbanna, Southwest China. *Chin. J. Ecol.* **2009**, *28*, 1942–1948.

The Rubber Board. Rubber Growers Companion. Government of India, Kottayam, Kerala, India, 115, 2005.

Thomas, S. C. Asymptotic Height as a Predictor of Growth and Allometric Characteristics in Malaysian Rain Forest Trees. *Am. J. Bot.* **1996**, *83*, 556–566.

UNFCCC. Kyoto Protocol to the United Nations Framework Convention on Climate Change. Adapted at COP3 in Kyoto, Japan, on December 11, 1997.

Upadhaya, K.; Thapa, N.; Barik, S. K. Tree Diversity and Biomass of Tropical Forests Under Two Management Regimes in Garo Hills of North-Eastern India. *Trop. Ecol.* **2015**, *56*, 257–268.

Wauters, J. B.; Coudert, S.; Grallien, E.; Jonard, M.; Ponette, Q. Carbon Stock in Rubber Tree Plantations in Western Ghana and MatoGrosso (Brazil). *For. Ecol. Manag.* **2008**, *255*, 2347–2361.

Whittaker, R. H.; Niering, W. A. Vegetation of the Santa Catalina Mountains, Arizona. V. Biomass, Production, and Diversity Along the Elevation Gradient. *Ecology* **1975**, *56*, 771–790.

World Rainforest Movement. REDD in Congo Basin, 2010. http://www.redd-monitor.org/wp-content/uploads/2010/12/REDD_Congo.pdf (accessed Jan 27, 2018).

Yang, J. C.; Huang, J. H.; Tang, J. W.; Pan, Q. M.; Han, X. G. Carbon Sequestration in Rubber Tree Plantations Established on Former Arable Lands in Xishuangbanna, SW China. *Acta Phytoecologica Sinica.* **2005**, *29*, 296–303.

Yang, X.; Blagodatskiy, S.; Cadisch, G.; Chu Xu, J. Carbon Storage Potential of Rubber Plantations of Different Age and Elevation in Xishuangbanna. Presented in "Bridging the Gap Between Increasing Knowledge and Decreasing Resources," Tropentag, September, Prague, Czech Republic, 2014; pp 17–19.

Zeide, B. Analysis of Growth Equations. *For. Sci.* **1993**, *39*, 594–616.

Ziegler, A. D.; Fox, J. M.; Xu, J. The Rubber Juggernaut. *Science* **2009**, *324*, 1024–1025.

CHAPTER 3

Potential Loss of Biomass Carbon

3.1 INTRODUCTION

Biomass stock or carbon accumulation rate in any tree depends on the growth rate of the tree, which is limited by various factors. Climatic variations, clone variations, planting space, and different management systems also affect biomass and carbon accumulation rate of rubber trees (Egbe et al., 2012; Mohammed, 2012; Harmon and Marks, 2002; Pussinen et al., 2002; Liski et al., 2001; Cooper, 1983). The growth rate, biomass density as well as carbon density of rubber trees are influenced by planting space between two rubber trees. The 2 m × 4 m spaced plantations, stocks 13.94%, 14.44% more biomass and 13.69%, 14.46% more carbon than 2 m × 3 m and 2 m × 2 m spaced plantations respectively in Ethiopia (Mohammed, 2012). Apart from the planting space different management systems with different rotation length have been considered as an effective management practice to manage the carbon stocks under rubber plantations (Harmon and Marks, 2002; Pussinen et al., 2002; Liski et al., 2001; Cooper, 1983). Rotation systems like managed extension and complete rotations directly influencing the carbon stocks in rubber plantation. Carbon stock increases with the increase of rotation length (Liski et al., 2001).

Plantations of the rubber tree currently occupy ~10 million ha land in the tropical parts of the world (FAO, 2010). Although traditionally grown on large estates, over the last century the cultivation of rubber has gradually moved to the small-holder sector which now accounts for more than 75% of the world's natural rubber production (Rodrigo et al., 2005; Ahrends et al., 2015). Rubber plantations are expanding rapidly throughout Asia, and the vast majority of these new plantations are monocultures as opposed to the traditional mixed rubber agroforestry systems (Ahrends et al., 2015; Villamor et al., 2014). More than 1.5 million ha of land has been converted into monoculture rubber plantations in Southeast Asia alone (Ziegler et al., 2009; Tan et al., 2011; de Blecourt et al., 2013). In India, 0.76 million ha of land is currently under rubber plantation and the annual increase in acreage is 3% (The Rubber Board, 2013). The small-holder sector in India represents 91% of the total area under rubber and 94% of the total natural rubber production (The Rubber Board, 2013) implying the significance of small-holder farming in land use changes.

The risks of this land-use change may be high as rubber plantations have often been associated with various environmental issues including biodiversity loss (Ahrends et al., 2015; Warren-Thomas et al., 2015), disruption of hydrological processes (Ziegler et al., 2011; Tan et al., 2011; Qui, 2009), and soil carbon (C) loss (de Blecourt et al., 2013). The conversion of forests with high biodiversity value to rubber plantations and expansion into marginal areas creates further opportunities for biodiversity loss (Ahrends et al., 2015; Warren-Thomas et al., 2015) and disruption of nutrient cycles following clear-felling. For example, between 2005 and 2010 over 512 km^2 of land in Key Biodiversity Areas, 610 km^2 in protected areas and 1624 km^2 in conservation corridors was converted into rubber plantations (Ahrends et al., 2015). Moreover, rubber plantations are also much bigger water consumers than rain forests and cause local water shortages (Tan et al., 2011). Conversion of secondary forests into rubber plantations was also shown to lead to soil C loss by about 19% of the initial stock (de Blecourt et al., 2013; Blagodatsky et al., 2016).

Rubber trees have been principally managed for commercial latex production up to their productive age followed by clear-felling and replanting (Egbe et al., 2012). The productive life of rubber plantations is approximately 40 years but productivity starts to decline after about 30 years (Egbe et al., 2012; FAO, 2001). Thus, currently the common rotation length is 30–35 years (Nizami et al., 2014). However, the nonuse values of these plantations, especially C sequestration, have not been fully appreciated. Despite their negative environmental impacts mentioned above, rubber plantations have immense potential for terrestrial C sequestration by accumulating C in woody biomass and organic matter in mineral soil (Egbe et al., 2012; Cheng et al., 2005; Corpuz et al., 2014; Sone et al., 2014). This can play an important role in mitigating climate change. However, the amount of C sequestered in woody biomass is very much dependent on plantation age, stem density, site conditions, management practices, and climatic condition (Egbe et al., 2012, Corpuz et al., 2014, Orjuela-Chaves et al., 2014).

In many places information about the biomass or C sequestration of rubber plantations is still scanty (Sone et al., 2014). Clear-felling of trees at the end of plantations may result in significant losses in system C (Zhou et al., 2013; Talberth et al., 2015; Egnell and Leijon, 1999; Egnell and Valinger, 2003). However, such losses have not been systematically quantified in rubber plantations, and many workers (Tan et al., 2011) have recommended further studies on C release. Therefore, it is important to estimate the biomass C stocks in plantations and biomass C lost through clear-felling.

Various destructive and non-destructive methods have been recommended for biomass estimation. Out of these methods, use of allometric model is a reliable and non-destructive means for assessing tree biomass (Kuyah et al., 2012; Goodman et al., 2014). Species- and site-specific biomass estimation models are preferred over generalized models because the former give more accurate predictions (Goodman et al., 2014; Litton and Kauffman, 2008; Chave et al., 2014). Species-specific models are usually developed using measurements of a single species at a particular site or across its range of distribution. Generalized models are developed using large global datasets from mixture of species across a broad range of conditions. Therefore, the use of generalized models can lead to a bias in estimating biomass for a particular species. Although species-specific models are available for rubber elsewhere (Blagodatsky et al., 2016) such models are lacking for India. We also did not want to entirely depend on published models because such models are usually based on a small number of directly harvested trees and include very few large diameter trees (Chave et al., 2005), thus may not be suitable for application outside the area and range of tree size (Sileshi, 2014). Therefore, the objectives of this work are to (1) develop and test models for estimating rubber tree biomass and C storage in situations where diameter at breast height (D 1.3 m) cannot be measured precisely due to the tapping artifact and (2) estimate losses in biomass C due to clear-felling of mature plantation. In this study, our focus was on estimation of biomass C losses after clear-felling and estimating changes in soil C was beyond the scope of this study. This is because, it is often difficult to distinguish such changes from the large spatial variability in soil C unless repeated measurements are taken over a long period after clearing the plantation.

3.2 METHODS FOR DATA COLLECTION

3.2.1 *Stand Selection and Sampling*

In order to estimate the biomass stocks of rubber plantations, five quadrates of 25 m × 25 m area were selected from a single rubber plantation of 35–40 years old. Since dead and unhealthy saplings were replaced by new saplings during the initial stage (≤4 years old) of rubber plantations, a stand of 40 years old may consists of trees that are 35–40 years old. The mean stem density of the stand was 714 ha^{-1}; with a range of 672–776 ha^{-1}. Measurements of total tree height and circumference at 2 m above the ground were recorded for a sample of live trees within the selected stands. Although the standard practice for measuring tree diameter is at breast height (D at 1.3 m; Brokaw and Thompson, 2000), in our case D could not be measured at 1.3 m because tree stems at this height were heavily deformed due to tapping for latex collection. Therefore, the circumference of the stems at 2 m above the ground level was measured to avoid such artifact. During circumference measurement of the standing trees the following recommended practices (Husch et al., 2003) were followed: (1) circumference was measured carefully on the uphill side of trees when trees were found on slopes or uneven ground, (2) circumference was measured parallel to the lean on the high side of the tree where the trees were found inclined, (3) circumference of each stem was measured when the tree consisted of two or more stems forking below the circumference measurement level of the tree, and (4) circumference was taken just below the enlargement of the stem when forking occurred at or above the circumference measurement level of the tree. Then representative diameter classes were developed based on stem circumferences as follows: 61–70, 71–80, 81–90, 91–100, 101–110 and 111–120 cm, and 9 trees were felled from each circumference class.

A total of 54 trees thus felled were used for development of biomass estimation models. The leaf, branch and stem parts were separated and total fresh weights (kg) of each plant part (PFW) were recorded for the felled trees. For belowground biomass (BGB), coarse root (>2.5 mm diameter) were extracted from a 1 m radius with 1 m depth around the felled tree stem. However, our study did not incorporate fine root biomass due to its difficulty in separation and sample collection. Fresh weights of extracted coarse roots were weighed after carefully removing the soil.

Sub-sample (500 g) of each plant part was taken to the laboratory and oven dried at 65°C for 72 h. Fresh weight and dry weight ratios (R) for each of the sub-sampled plant part were calculated after estimating their respective dry weight. Dry weight of each part of the sampled tree (PDW) was estimated as PDW = (R × PFW). Total biomass (B) was obtained by adding all the calculated parts of the respective tree, that is, as the sum of aboveground biomass (AGB) and BGB.

3.2.2 Estimation of Tree Biomass and C Densities

Models based on diameter at breast height (D at 1.3 m) alone or including total height (H) or/and wood density (ρ) are widely used in predicting tree biomass (Goodman et al., 2014; Chave et al., 2014; Sileshi, 2014). In this study, we first tested two types of species-specific models, namely, (1) simple power-law (allometric) models that require either D or H as the only predictor and (2) models that consist of compound variables. The simple power-law models have been demonstrated empirically and supported by emergent theories of macroecology (Sileshi, 2014). Different theoretical models derived from physical and biological first principles predict universal scaling relationships between total tree biomass (B), H, and D (Sileshi, 2014). For this study, we explored allometric relationships between D at 2 m (hereafter D_2) and H expressing the relationships as follows:

$$\ln(B) = \ln(\alpha) + \beta(lnH) + \varepsilon \qquad \text{Model 1}$$

$$\ln(B) = \ln(\alpha) + \beta(\ln D^2) + \varepsilon \qquad \text{Model 2,}$$

where ε is the error. Back-transformation of Models 1 and 2 gives a power function as $B = \alpha X^\beta \times CF$ where X is either H or D and CF is the correction factor calculated from the mean square of error (MSE) as exp (*MSE*/2).

We also tested the performance of models that consists of compound variables. These models consist of various combinations of D, H and use wood specific gravity (ρ). For our purpose, we chose the following two models as they have been used recently in estimating biomass of rubber trees (Sone et al., 2014) and other tree species (Chave et al., 2005; Sileshi, 2014):

$$\ln(B) = \ln(\alpha) + \beta\left(\ln\left(HD^2\right)\right) + \varepsilon \qquad \text{Model 3}$$

$$B = \alpha\left(\rho D^2 H\right)^{\beta} \qquad \text{Model 4.}$$

In addition, we tested the performance of the general pan tropical model below:

$$B = 0.0673\left(\rho D^2 H\right)^{0.976} \qquad \text{Model 5.}$$

This model has been proposed to provide estimates of biomass for broadly defined forest types (Chave et al., 2014). In Model 4, the IPCC value of wood density (ρ) for rubber in Asia (0.53) was used. The values of D in Models 2–4 were calculated from the circumference measured at 2 m. Therefore, throughout this paper D means stem diameter at 2 m (unlike the usual 1.3 m). Parameters of Models 1–4 were estimated directly from the data using the SAS system while those of Model 5 are directly obtained from Chave et al. (2014).

We compared the performance of Models 1–4 using the coefficient of determination (R^2), the prediction error (Goodman et al., 2014; Chave et al., 2014), Akaike information criterion (AIC) and the mean absolute percentage error (MAPE) as proposed by Sileshi (2014). Inferences using the prediction error and MAPE were all based on the estimates and their 95% confidence limits (CL) because in situations like this, the 95% confidence interval (CI) serves as a conservative test of hypothesis and it also attaches a measure of uncertainty to sample statistic (Sim and Reid, 1999). For comparing the range of models being tested we used the Akaike weights as they are more informative than AIC when in model comparison (Sileshi, 2014). However, all of the criteria above do not tell us whether the estimated biomass will be significantly different from the measured biomass or not in future samples. This is a very important consideration because maximizing model fit may not necessarily translate into biologically meaningful differences in biomass estimates. Therefore, we compared the means and 95% CI of the measured tree biomass with predictions from the different models. Measured and predicted means were considered to be significantly different from one another only if their 95% CI were nonoverlapping.

Biomass density (Mg ha^{-1}) was estimated using the following formula:
Biomass density = $N/n \times (B_1 + B_2 + B_3 + B_4 + \ldots\ldots + B_n)$,
where N is the stem density, n is the number of trees harvested, and B_i is the biomass stock (kg) of the respective tree. The mean stem density of the stand was 714 ha^{-1}. To estimate the amount of C sequestered by the plantations, biomass was converted into C by multiplying by 0.5, which is a factor proposed by IPCC (IPCC, 2008). The results were then converted to a per hectare basis. The main C pools applicable to land-use, land-use change and forestry activities are aboveground trees, aboveground non-tree, belowground roots, litter, dead wood, and soil organic matter. In this study, we focused on the C pool in the above and belowground living biomass (excluding fine roots) of trees since these account for the largest percentage of the sequestered C within a forest ecosystem, and therefore the most important C pool in the land-use change context (Kongsager et al., 2013). Other C pools such as litter, dead wood and soil organic matter, were not considered here. Total biomass C densities, as well as C, in the aboveground and belowground compartments were estimated. The C accumulation rates were estimated by dividing the total biomass C by the tree age. The C loss due to clear-felling was inferred from the C storage assuming that all C stored in AGB will be lost from the system. This is because the common practice following clear-felling is the removal of all the timber and burning of the residues when preparing the land for replanting.

3.3 BIOMASS ESTIMATION MODELS

Strong allometric relationships were found between total tree height (H) and diameter at 2 m (D) (Fig. 3.1a), BGB and H (Fig. 3.1b), AGB and H (Fig. 3.1c), AGB and D (Fig. 3.1d), and BGB and D (Fig. 3.1e). The estimated coefficients of Models 1–4 are summarized in Table 3.1. According to the measures of model goodness of fit, Models 1–4 are indistinguishable, but all gave predictions with lower errors compared to Models 5 (Table 3.2). Models 3 and 4 had the largest AIC weights indicating that these two models have better fit than Models 1 and 2. However, Models 3 and 4 are indistinguishable with all criteria.

TABLE 3.1 Coefficients of the Site-specific Biomass Estimation Models.

Model	Variable	Intercept	Exponent	RMSE
1	H	0.0007 (0.0002–0.002)	4.19 (3.84–4.54)	0.097
2	D	1.31 (0.79–2.17)	1.71 (1.56–1.86)	0.103
3	HD^2	0.34 (0.19–0.60)	0.72 (0.66–0.78)	0.096
4	$HD^2\rho$	0.53 (0.31–0.91)	0.72 (0.66–0.78)	0.096

Note: Figures in parenthesis are the 95% confidence limits.
ρ, wood density for rubber in Asia; D, diameter of the tree at 200 cm height; H, total height of the tree; RMSE, root mean square error.

The total biomass predicted by Models 1–4 was also not significantly different from the measured biomass (Fig. 3.2a,c); the predicted tree biomass being within 0.2% and 5% of the measured biomass. On the other hand, the general model (Model 5) significantly overestimated total biomass compared to the measured biomass; its error of prediction being over 35% (Table 3.2). The allometric model consisting of H as the sole predictor underestimated total tree biomass by 3.8%, while biomass predictions from models consisting of D alone or D, H, and ρ were indistinguishable from the measured biomass (Table 3.2). Indeed, the prediction from the model that consists of all the variables (D, H, and ρ) was the same as the model consisting of only D or H (Fig. 3.2b).

TABLE 3.2 Comparison of Predictions of the Site Specific Models and General Model Using Goodness of Fit Criteria.

Model	Model	Variable	R^2	Error (%)	MAPE	AIC	AICw
Local	1	H	0.918	−3.8 (−6.3–1.2)	7.7 (5.9–9.4)	−249.7	0.161
	2	D	0.908	0.5 (−2.2–3.3)	9.1 (7.9–10.2)	−243.2	0.006
	3	HD^2	0.921	0.4 (−2.1–2.9)	8.1 (6.9–9.3)	−251.6	0.416
	4	$HD^2\rho$	0.921	0.4 ((−2.1–2.9)	8.1 (7.0–9.3)	−251.6	0.416
General	5	$HD^2\rho$	NA	35.2 (29.3–41.0)	35.2 (29.3–41.0)	NA	NA

Note: Figures in parentheses are the 95% confidence limits.
ρ, wood density for rubber in Asia; AIC, Akaike information criterion; AICw, Akaike weights; D, diameter of the tree at 200 cm height; H, total height of the tree; MAPE, mean absolute percentage error; R^2, coefficient of determination.

Potential Loss of Biomass Carbon

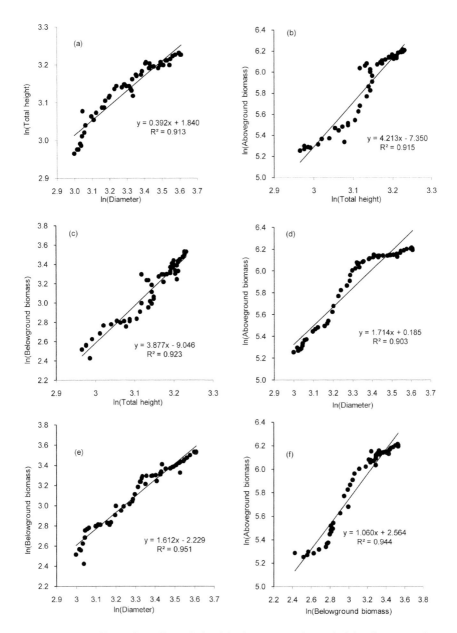

FIGURE 3.1 Allometric scaling relationships between total stem height, diameter at 2 m (D_2), aboveground and belowground tree biomass.

3.4 TREE BIOMASS, C STORAGE, AND SUBSEQUENT LOSS

The mean aboveground tree biomass was 376.9 kg, while BGB was 23.9 kg (Table 3.3). With a scaling exponent of 1.06 (95% CL: 0.99–1.13) the relationship between the aboveground and belowground tree biomass was isometric rather than allometric (Fig. 3.1f). Tree biomass significantly differed among circumference classes; trees with circumferences of 60–80 cm having lower and more variable biomass than those with circumferences of 80 cm or more (Fig. 3.2b). The C densities predicted by our models were also very close to C densities calculated from measured tree biomass (Fig. 3.2d).

On an area basis, the estimated total biomass was 286.2 Mg ha^{-1}, out of which 269.1 Mg ha^{-1} was in aboveground and 17 Mg ha^{-1} in belowground parts (Table 3.3). The estimated total C density in tree biomass was 143.1 Mg ha^{-1}; 134.6 Mg ha^{-1} being in aboveground and 8.5 Mg ha^{-1} in belowground parts (Table 3.3). The C accumulation rate was estimated at 3.6 Mg ha^{-1} y^{-1} with a minimum and maximum of 1.8 and 4.8 Mg ha^{-1} y^{-1}, respectively (Table 3.3). The C loss due to clear-felling of mature plantations was estimated at 135 Mg ha^{-1}; the minimum and maximum C losses being 68.2 and 178 Mg ha^{-1}, respectively (Table 3.3).

TABLE 3.3 The Mean and Median Biomass C Densities in Rubber Plantations in Karimganj District, State of Assam in North East India, and Potential C Losses due to Clear-Felling. All trees are around 35–40 years old.

Variable	Tree part	Mean	Median
Tree biomass (kg tree^{-1})	Aboveground	376.9 (346.7–407.1)	434.9
	Belowground	23.9 (22.0–25.7)	25.5
	Total	400.8 (368.8–432.8)	460.6
Biomass density (Mg ha^{-1})	Aboveground	269.1 (247.6–290.7)	310.5
	Belowground	17.0 (15.7–18.4)	18.2
	Total	286.2 (263.3–309.0)	328.9
C density (Mg ha^{-1})	Aboveground	134.6 (123.8–145.3)	155.3
	Belowground	8.5 (7.9–9.2)	9.1
	Total	143.1 (131.7–154.4)	164.5
C accumulation rate (Mg ha^{-1} y^{-1})	Aboveground	3.4 (3.1–3.6)	3.9
	Belowground	0.21 (0.20–0.23)	0.2
	Total	3.6 (3.3–3.9)	4.1
C loss (Mg ha^{-1})	Aboveground	134.6 (123.8–145.3)	155.3

Note: Values in parenthesis are 95% confidence limits. Results expressed as mean value and ranges are given in the parenthesis.

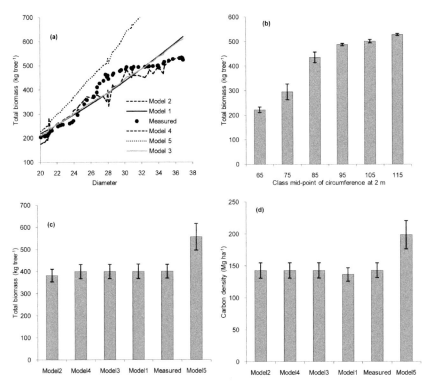

FIGURE 3.2 **(See color insert.)** Comparison of (a) trends in measured (solid circles) with the predicted total tree biomass, (b) variations in tree biomass with stem circumference, (c) measured tree biomass, and (d) carbon density with predictions from different models. Vertical bars represent 95% confidence limits of estimates.

3.5 BIOMASS STOCKS AND POTENTIAL BIOMASS STOCKS LOSS

Although the traditional practice when developing biomass estimation models has been to use D at 1.3 m and other variables, in the tree stands we studied it was not possible to measure D at 1.3 m due to the tapping artifact. Therefore, in this work we tested and develop biomass estimation models that fit our study condition. We found strong allometric relationships between tree biomass and D at 2 m and total tree height. We also found isometric relationship between AGB and BGB, which was consistent with allometry theory (Enquist and Niklas, 2002) and empirical observations (Cheng and Niklas, 2007; Hui et al., 2014).

In this study, models containing either H or D alone predicted tree biomass as accurately as models with compound variables. Comparison of Models 3 and 4 indicate that the inclusion of wood density did not improve biomass prediction. The findings also suggest that D measured at 2 m may be as good as D 1.3 m. This has significant implications especially in situations where D 1.3 m cannot be measured precisely due to the tapping artifact. However, for studies to be comparable and consistent across sites we still believe that D should be measured at 1.3 m in rubber plantations wherever possible. This is because the ambiguity with which D was defined in the past has been a key source of error in biomass estimates (Brokaw and Thompson, 2000) and allometry coefficients (Sileshi, 2014, 2015). Indeed, small differences in D measurement can lead to large discrepancies in biomass estimates. According to Brokaw and Thompson (2000), biomass estimates were 4% higher when calculated using D measured at 1.3 m than at 1.4 m. Since the prediction from Models 1–4 did not significantly differ from each other and the measured values, any of our Models (1–3) may be used in plantations of similar ages and circumference distribution. However, it is important to sample more trees with circumferences of <80 cm to account for the higher variability observed in biomass (Fig. 3.2b). Smaller trees (<80 cm) may be more variable probably because they are likely to be unhealthy or saplings suppressed by bigger trees during the establishment phase.

The total biomass of 286 Mg ha^{-1} found in this study is comparable with biomass stocks reported indifferent agroforestry (Kirby and Potvin, 2007; Kumar and Nair, 2011) and forestry systems and plantations (Mandal and Islam, 2008) across the world. In India, the biomass stock reported for 30-year-old rubber plantation was 191.6 Mg ha^{-1} (Mandal and Islam, 2011). Similarly, biomass stock of 211.9 Mg ha^{-1} in 30-year-old rubber plantation was reported from Sri Lanka (Munasinghe and Rodrigo, 2014). In the Philippines, a 35 years rubber plantation had total biomass of 246.2 Mg ha^{-1} (Corpuz et al., 2014). Our estimate of C stocks (143 Mg ha^{-1}) is also comparable with values reported in the literature. The mean total (above + belowground) C stock ranges from 16–106 Mg ha^{-1} under 5–40-year-old rubber plantations (Brahma et al., 2016). In a 35-year-old rubber plantation in the Philippines C density was estimated at 110.8 Mg ha^{-1} (Corpuz et al., 2014). In Ghana, C stored in ~40-year-old rubber plantation was 213.6 Mg ha^{-1} (Kongsager et al., 2013). The C accumulation rate (3.6 Mg ha^{-1}y^{-1}) recorded in this study is close to those reported for plantations of comparable age. For example, in Ghana, the C accumulation rate was 4.9 Mg ha^{-1}y^{-1} in ~40-year-old rubber plantation (Kongsager et al., 2013).

Although mature rubber plantations stock substantial amounts of biomass C they are prone to degradation with the existing clear-felling management system. According to our estimates, clear-felling of trees can lead to C loss of 68–178 Mg ha^{-1} from the system even if only aboveground tree biomass is removed. These figures could underestimate the actual losses because we have not considered other C pools such as litter, dead wood, and soil C. However, they are within the range of C losses reported following logging of plantations. For example, when logged forests are converted to continuous cropping, 90–200 Mg ha^{-1} C are lost from aboveground and 25 Mg ha^{-1} C are lost from the top soil (Hairiah et al., 2011). Losses of C from the conversion of logged forests to other tree-based systems are smaller, and range between 40 and 180 Mg ha^{-1} from aboveground and 10 Mg ha^{-1} from the soil (Hairiah et al., 2011).

Although latex production and C stocks increase with increasing rotation length, there are rotation lengths that optimize both economic and ecological benefits (Nizami et al., 2014; Song et al., 2013). Earlier studies on C inputs and latex production indicate that 40 years rotations are best suited for rubber plantation (Nizami et al., 2014; Song et al., 2013). For example, latex production reached a maximum (4.4 Mg ha^{-1} y^{-1}) at 40 years compared to production (3.5 Mg ha^{-1} y^{-1}) at the currently adopted rotation length of 35 years in China (Nizami et al., 2014). In order to conserve biomass, C stock in the plantations and to maintain ecosystem integrity we recommend selective felling of 20% of the growing stock starting from plantation age of 35 years and replanting at two-year cycles. Adopting the two-year felling cycle will take ten years for complete removal of mature trees and simultaneously over this period the newly planted trees will start commercial latex production. The proposed felling cycle is expected to maintain steady vegetative C stocks as well as continuous latex production. Alternatively, promotion of rubber agroforestry could be a strategy to combat system C losses. Elsewhere rubber agroforestry systems have been known to support relatively higher biodiversity, greater C stocks and local livelihoods although they may have relatively low latex productivity compared to monoculture plantations (Villamor et al., 2014). Promotion of diversified agroforestry systems in which rubber plays an important role, but is not planted as monocultures, may be more beneficial (Ziegler et al.,

2009; Nizami et al., 2014). In agroforestry arrangement, the introduction of economically and ecologically important species should be carried out in between the rubber trees according to the suitability from the age of 35 years.

One of the limitations of our study is our inability to estimate soil C losses and associated changes due to conversion of forests to rubber plantation and subsequent clear-felling. Conversion of forests to rubber plantations has been shown to result in significant decline in soil C stocks (de Blecourt et al., 2013; Blagodatsky et al., 2016; Guillaume et al., 2015). For example, in Yunnan province of China soil C decline of up to 80% of the original C was recorded (de Blecourt et al., 2013). Information is scanty on the long-term effect of clear-felling rubber on the soil C cycle and loss of soil C stocks. Clear-felling may lead to rapid turnover of detrital and soil C pools, including roots, litter and mineral soil organic matter (Lacroix et al., 2016) leading to losses over several years. Therefore, estimation of such loss in rubber plantations should be a subject of future studies.

3.6 CONCLUSIONS

The following conclusions emerged from this analysis: (1) the power-law model based on either D or H predicted biomass stocks in rubber as accurately as models consisting of compound variables (D, H, and ρ), (2) mature rubber plantations hold substantial C (on average 143 Mg ha^{-1}) in BGB and AGB, and (3) clear-felling of trees can lead to loss of all of the C stored in AGB (on average 135 Mg ha^{-1}) from the system. Therefore, best management practices need to be developed in order to minimize system C losses.

KEYWORDS

- **allometric models**
- **H-D models**
- **biomass density**
- **coarse root**
- **ecological stability**

REFERENCES

Ahrends, A.; Hollingsworth, P. M.; Ziegler, A. D.; Fox, J. M.; Chen, H.; Su, Y.; Xu, J. Current Trends of Rubber Plantation Expansions May Threaten Biodiversity and Livelihoods. *Glob. Environ. Chang.* **2015**, *34*, 48–58.

Blagodatsky,S.; Xu. J.; Cadisch, G. Carbon Balance of Rubber (*Hevea brasiliensis*) Plantations: A Review of Uncertainties at Plot, Landscape and Production Level. *Agric. Ecosyst. Environ.* **2016**, *221*, 8–19.

Brahma, B.; Nath, A. J.; Das, A. K. Managing Rubber Plantations for Advancing Climate Change Mitigation Strategy. *Curr. Sci.* **2016**, *110*, 2015–2019.

Brahma, B.; Nath, A. J.; Sileshi, G. W.; Das, A. K.; Estimating Biomass Stocks and Potential Loss of Biomass Carbon Through Clear-Felling of Rubber Plantations, Biomass and Bioenergy, 115, 2018, 88–96.

Brokaw, N.; Thompson, J. The H for DBH. *For. Ecol. Manag.* **2000**, *129*, 89–91.

Chave, J.; Andalo, C.; Brown, S.; Cairns, M.; Chambers, J.; Eamus, D. et al. Tree Allometry and Improved Estimation of Carbon Stocks and Balance in Tropical Forests. *Oecologia* **2005**, *145*, 87–99.

Chave, J.; Réjou-Méchain, M. E.; Búrquez, A. et al. Improved Allometric Models to Estimate the Aboveground Biomass of Tropical Trees. *Glob. Chang. Biol.* **2014**, *20*, 3177–3190.

Cheng, D.; Niklas, K. Above- and Below-Ground Biomass Relationships Across 1543 Forested Communities. *Ann. Bot.* **2007**, *99*, 95–102.

Cheng, Y. J.; Huang, J. H.; Tang, J. W.; Pan. Q. M.; Han, X. G. Carbon Sequestration in Rubber Tree Plantations Established on Former Arable Lands in Xishuangbanna, SW China. *Acta Phytoecol. Sin.* **2005**, *29*, 296–303.

Cooper, C. F. Carbon Storage in Managed Forests. *Can. J. For. Res.* **1983**, *13*, 155–166.

Corpuz, O. S.; Abas, E. L.; Salibio, F. C. Potential Carbon Storage of Rubber Plantations. *Ind. J. Pharm. Biol. Res.* **2014**, *2*, 73–82.

de Blécourt, M.; Brumme, R.; Xu, J.; Corre, M. D.; Veldkamp, E. Soil Carbon Stocks Decrease Following Conversion of Secondary Forests to Rubber (*Hevea brasiliensis*) Plantations. *PLoS One* **2013**, *8*, e69357.

Egbe, A. E.; Tabot, P. T.; Fonge, B. A.; Becham, E. Simulation of the Impacts of Three Management Regimes on Carbon Sinks in Rubber and Oil Palm Plantation Ecosystems of Southwestern Cameroon. *J. Ecol. Nat. Environ.* **2012**, *46*, 154–162.

Egnell, G.; Leijon, B. Survival and Growth of Planted Seedlings of *Pinus sylvestris* and *Picea abies* After Different Levels of Biomass Removal in Clear-felling. *Scandinavian J. For. Res.* **1999**, *14*, 4, 303–311.

Egnell, G.; Valinger, E. Survival, Growth, and Growth Allocation of Planted Scots Pine Trees After Different Levels of Biomass Removal in Clear-Felling. *For. Ecol. Manag.* **2003**, *177*, 65–74.

Enquist, B. J.; Niklas, K. J. Global Allocation Rules for Patterns of Biomass Partitioning in Seed Plants. *Science.* **2002**, *295*, 1517–1520.

FAO. Non-forest tree plantations, Report based on the work of Killmann, W. Forest Plantation Thematic Papers, Working Paper 6; Forest Resources Development Service,

Forest Resources Division, FAO, Rome, 2001. ftp://ftp.fao.org/docrep/fao/006/ac126e/ac126e00.pdf (accessed Dec 10, 2011).

FAO. *Global Forest Resource Assessment 2010. Food and Agricultural Organization of The United Nations (FAO)*, Rome, 2010.

Goodman, R. C.; Phillips, O. L.; Baker, T. R. The Importance of Crown Dimensions to Improve Tropical Tree Biomass Estimates *Ecol. Appl.* **2014**, *24*, 680–698.

Guillaume, T.; Damris, M.; Kuzyakov, Y. Losses of Soil Carbon by Converting Tropical Forest to Plantations: Erosion and Decomposition Estimated by $\delta 13C$. *Glob. Chang. Biolog.* **2015**, *21*, 3548–3560.

Hairiah, K.; Dewi, S.; Agus, F.; Velarde, S.; Ekadinata, A.; Rahayu, S. et al. *Measuring Carbon Stocks Across Land Use Systems: A Manual*. World Agroforestry Centre (ICRAF), SEA Regional Office, Bogor, Indonesia, 2011. http://www.worldagroforestry.org/sea/Publications/files/manual/M N0050-11/MN0050-11-1.pdf (accessed Apr 21, 2016).

Harmon, M. E.; Marks, B. Effects of Silvicultural Practices on Carbon Stores in Douglas-fir-Western Hemlock Forests in the Pacific Northwest, USA: Results from a Simulation Model. *Can. J. For. Res.* **2002**, *32*, 863–877.

Hui, D.; Wang, J.; Shen, W.; Le, X.; Ganter, P.; Ren, H. Near Isometric Biomass Partitioning in Forest Ecosystems of China. *PLoS One* **2014**, *9*, e86550.

Husch, B.; Beers, T. W.; Kershaw, J. A. *Forest Mensuration*, 4th ed.; John Wiley & Sons, Inc.: Hoboken, New Jersey, 2003; Vol. 85, p. 86.

IPCC, 2008. Climate change 2007, *Synthesis Report, Contribution of Working Groups I, II and III to the Fourth Assessment Report. In Intergovernmental Panel on Climate Change (IPCC)*; Pachauri, R. K., Reisinger, A. Eds.; Geneva-Switzerland, p. 104. Available at http://bit.ly/17pJfjj (accessed Nov 13, 2013)

Kirby, K. R.; Potvin, C. *Variation in Carbon Storage Among Tree Species: Implications for the Management of a Small-Scale Carbon Sink Project (*Supplementary Material) *For. Ecol. Manag.* **2007**, *246*, 208–221.

Kongsager, R.; Napier, J.; Mertz, O. The Carbon Sequestration Potential of Tree Crop Plantations. *Mitigat. Adapt. Strat. Glob. Chang.* **2013**, *18*, 1197–1213.

Kumar, B. M.; Nair, P. K. R. Carbon Sequestration Potential of Agroforestry 3 Systems: Opportunities and Challenges. In *Advances in Agroforestry 8*; Springer, Dordrecht Heidelberg: London, New York, 2011.

Kuyah, S.; Dietz, J.; Muthuri, C.; Jamnadass, R.; Mwangi, P.; Coe, R.; Henry, N. Allometric Equations for Estimating Biomass in Agricultural Landscapes: I. Aboveground Biomass. *Agri. Ecosyst. Environ.* **2012**, *158*, 216–224.

Lacroix, E. M.; Petrenko, C. L.; Friedland, A. J. Evidence for Losses from Strongly Bound SOM Pools After Clear Cutting in a Northern Hardwood Forest. *Soil Sci.* **2016**, *181*, 202–207.

Liski, J.; Pussinen, A.; Pingoud, K.; Mäkipää, R.; Karjalainen, T. Which Rotation Length Is Favourable to Carbon Sequestration? *Can. J. For. Res.* **2001**, *31* (11), 2004–2013.

Litton, C. M.; Kauffman, J. B. Allometric Models for Predicting Aboveground Biomass in Two Widespread Woody Plants in Hawaii. *Biotropica* **2008**, *40* (3), 313–320.

Mandal, D.; Islam, K. R. Carbon Sequestration in Sub-Tropical Soils Under Rubber Plantations in North-East India. Paper presented at the International Symposium on Climate Change and Food Security in South Asia, Dhaka, Bangladesh, 2008. www.

wamis.org/agm/meetings/rsama08/Bari401-Islam-CarbonSequestration-Rubber.pdf (accessed Nov 15, 2014)..

Mohammed, A. Growth and Carbon Storage of Different Spaced Rubber Tree (*Hevea brasiliensis*) Plantation at Bebeka Coffee Farm, South-West Ethiopia. *Ethiop. J. Appl. Sci. Tech.* **2012**, *3* (2), 45–51.

Munasinghe, E. S.; Rodrigo, V. H. L.; Gunawardena, U. A. D. P. Modus Operandi in Assessing Biomass and Carbon in Rubber Plantations Under Varying Climatic Conditions. *Experiment. Agri.* **2014**, *50*, 40–58.

Nizami, S. M.; Yiping, Z.; Liqing, S. Zhao, W.; Zhang, X. Managing Carbon Sinks in Rubber (*Hevea brasilensis*) Plantation by Changing Rotation Length in SW China. *PLoS One* **2014**, *9*, e115234.

Orjuela-Chaves, J. A.; Andrade, C. H. J.; Vargas-Valenzuela, Y. Potential of Carbon Storage of Rubber (*Hevea brasiliensis* Müll. Arg.) Plantations in Monoculture and Agroforestry Systems in the Colombian Amazon. *Tropi. Subtrop. Agroecosyst.* **2014**, *17*, 231–240.

Pussinen, A.; Karjalainen, T.; Makipaa, R.; Valsta, L.; Kellomaki, S. Forest Carbon Sequestration and Harvests in Scots Pine Stand under Different Climate and Nitrogen Deposition Scenarios. *For. Ecol. Manag.* **2002**, *158*, 103–115.

Qiu, J. Where the Rubber Meets the Garden. *Nature.* **2009**, *457*, 246–247.

Rodrigo, V. H. L.; Stirling, C. M.; Silva, T. U. K.; Pathirana, P. D. The Growth and Yield of Rubber at Maturity is Improved by Intercropping with Banana During the Early Stage of Rubber Cultivation. *Field Crops Res.* **2005**, *91*, 23–33.

Sileshi, G. W. A Critical Review of Forest Biomass Estimation Models, Common Mistakes and Corrective Measures. *For. Ecol. Manag.* **2014**, *329*, 237–254.

Sileshi, G. W. The Relative Standard Error as an Easy Index for Checking the Reliability of Regression Coefficients, 2015. Doi: 10.13140/RG.2.1.2123.6968.

Sim, J.; Reid, N. Statistical Inference by Confidence Intervals: Issues of Interpretation and Utilization. *Phys. Therapy* **1999**, *79*, 186–195.

Sone, K.; Watanabe, N.; Takase, M.; Hosaka, T.; Gyokusen, K. Carbon Sequestration, Tree Biomass Growth and Rubber Yield of PB260 Clone of Rubber Tree (*Hevea brasiliensis*) in North Sumatra. *J. Rubb. Res.* **2014**, *17*, 115–127.

Song, Q. H.; Tan, Z.; Zhang, Y.; Sha, L.; Deng, X.; Deng, Y. et al. Do the Rubber Plantations in Tropical China Act as Large Carbon Sinks? *iForest-Biogeosci. For.* **2013**, *7*, 42–47.

Talberth, J.; Dellasala, D.; Fernandez, E. Global Forest Watch Report; GEOS Institute, Center for Sustainable Economy: Oregon, 2015.

Tan, Z.-H.; Zhang, Y. P; Song, Q. H.; Liu, W. J.; Deng, X. B.; Tang, J. W.; Deng, Y.; Zhou, W. J.; Yang, L. Y.; Yu, G. R.; Sun, X. M.; Liang, N. S. Rubber Plantations act as Water Pumps in Tropical China. *Geophys. Res. Lett.* **2011**, *38*, L24406.

The Rubber Board, 2013. Indian Rubber Statistics. The Statistics and planning Department, Ministry of Commerce and Industry, Govt. of India. Vol. 36, 2013. http://rubberboard.org.in/rubberstaticsdisplaypage.asp.

Villamor,G. B.; Le, Q. B.; Djanibekov, U.; van Noordwijk, M.; Vlek, P. L. G. Biodiversity in Rubber Agroforests, Carbon Emissions, and Rurallivelihoods: An Agent-Based Model of Land-use Dynamics in Lowlandsumatra. *Environ. Modell. Soft.* **2014**, *61*, 151–165.

Warren-Thomas, E.; Dolman, P. M.; Edwards, D. P. Increasing Demand for Natural Rubber Necessitates a Robust Sustainability Initiative to Mitigate Impacts on Tropical biodiversity. *Conserv. Lett.* **2015**, *8*, 230–241.

Zhou, D.; Liu, S.; Oeding, J.; Zhao, S. Forest Cutting and Impacts on Carbon in the Eastern United States. *Sci. Rep.* **2013,** *3,* 3547.

Ziegler, A. D.; Fox, J. M.; Xu, J. The Rubber Juggernaut. *Science.* **2009,** *324,* 1024–1025.

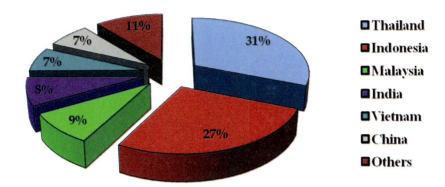

FIGURE 1.1 Share of natural rubber production in 2011.

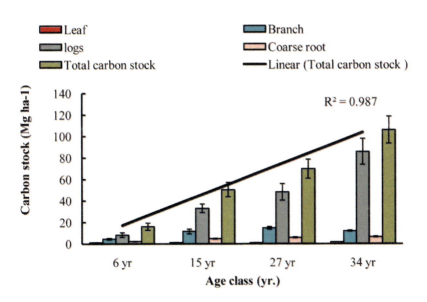

FIGURE 2.5 Measured Carbon stock (Mg ha^{-1}) under different aged rubber tree plantations.

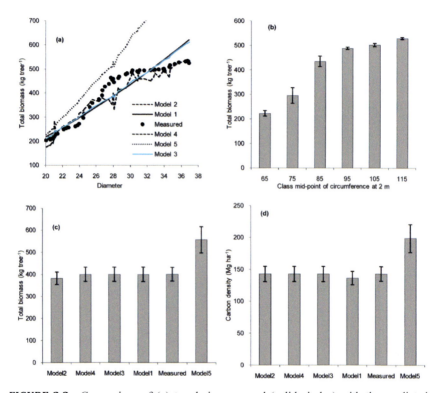

FIGURE 3.2 Comparison of (a) trends in measured (solid circles) with the predicted total tree biomass, (b) variations in tree biomass with stem circumference, (c) measured tree biomass, and (d) carbon density with predictions from different models. Vertical bars represent 95% confidence limits of estimates.

Rubber Plantations and Carbon Management

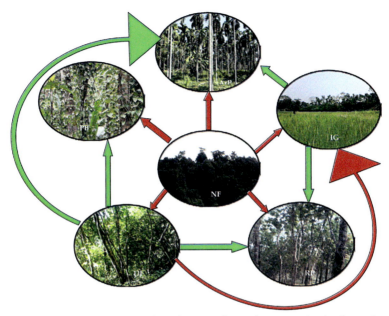

FIGURE 5.1 Schematic presentation of progressive and retrogressive land use changes. Green color arrow denotes progressive changes and red color retrogressive changes.

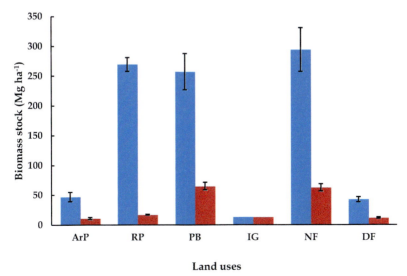

FIGURE 5.2 Biomass stock (above and below ground) for predominant land uses of northeast India. Values are mean ± standard error. Blue bars: aboveground biomass stock and red bars: belowground biomass stock. ArP, *Areca* plantation, DF, disturbed forest; IG, *Imperata* grassland; NF, natural forest; PB, *Piper betle* agroforestry; RP, rubber plantation.

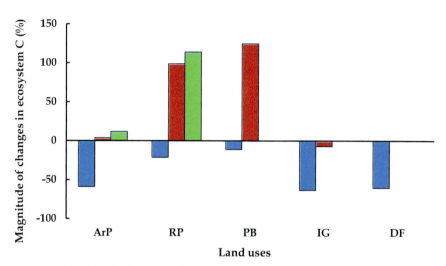

FIGURE 5.3 Magnitude (%) of changes in ecosystem C over their respective control land use. Blue bars: magnitude of ecosystem carbon changes considering NF as control, red bars: considering DF as control, and green bars: considering IG as control land use. ArP, *Areca* plantation; DF, disturbed forest; IG, *Imperata* grassland; PB, *Piper betle* agroforestry; RP, rubber plantation.

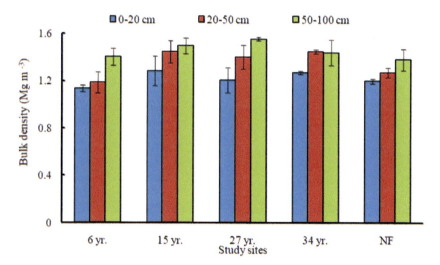

FIGURE 6.1 Soil bulk density (Mg m^{-3}) for different soil depths under different aged rubber plantations and natural forest. Error bar represents standard error of the mean. In the figure, 6, 15, 27, and 34 years are rubber tree plantations under different ages and NF represents the natural forest.

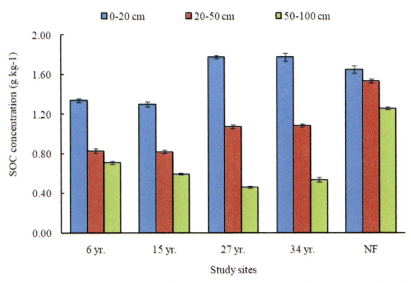

FIGURE 6.2 Soil organic carbon (SOC) concentration for different soil depth under different aged rubber plantations and natural forest. Error bar represents standard error of the mean. In the figure, 6, 15, 27, and 34 years are rubber tree plantations under different ages and NF represents the natural forest.

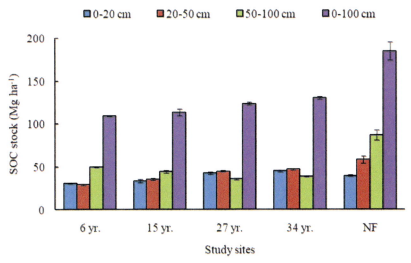

FIGURE 6.3 Soil organic carbon stock (Mg ha^{-1}) for different soil depths under different aged rubber plantations and natural forest. Error bar represents standard error of the mean. In the figure, 6, 15, 27, and 34 years are rubber tree plantations under different ages and NF represents the natural forest.

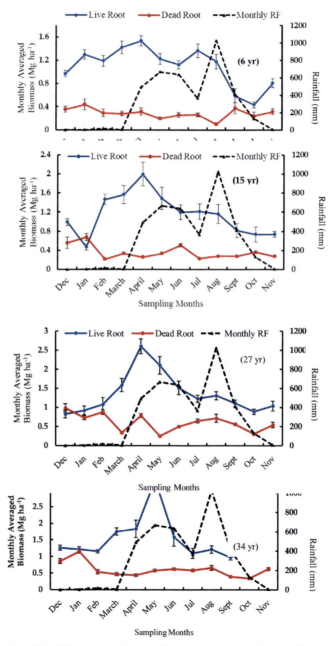

FIGURE 7.4 Monthly observed root biomass with cumulative RF under rubber plantations of Barak Valley.

Rubber Plantations and Carbon Management

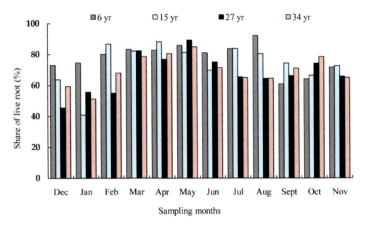

FIGURE 7.5 Monthly share of live root biomass to the monthly standing fine root biomass.

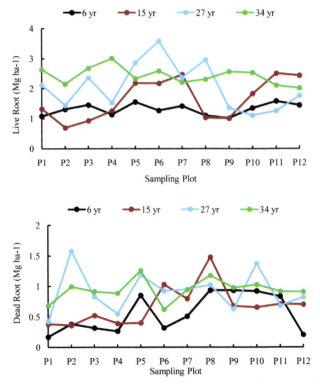

FIGURE 7.6 Estimated live and dead fine root biomass production under different ages of rubber plantations.

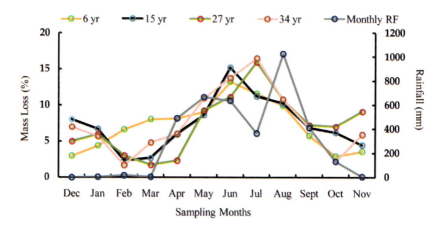

FIGURE 7.7 Estimated mass loss due to decomposition in different months over the study period.

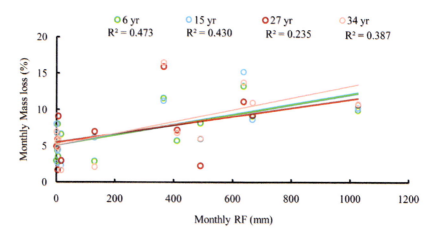

FIGURE 7.8 Coefficient of determination for the relationship between monthly mass loss and monthly rainfall.

CHAPTER 4

Impact of Land Use Changes on Soil Organic Carbon

4.1 INTRODUCTION

The global soil organic carbon (SOC) stock of ~1.5 × 10³ Pg (1 Pg = 10¹⁵ g) represents two and three folds higher than that of the atmosphere and vegetation, respectively (Jobbagy and Jackson, 2000; Lal, 2016). Tropical forests have been known for alteration of atmospheric C concentration by acting as C sink or source (Wei et al., 2014). Human-induced land use change (LUC) from tropical forests into plantations and croplands may act as a C source (De Blecourt et al., 2013; Guillaume et al., 2015; Fan et al., 2016; Iqbal and Tiwari, 2016) and simultaneously degrade soil properties (Abera and Wolde-Meskel, 2013). Of the total of 350 M ha of tropical forest land that has been converted to different land uses (ITTO, 2002), ~1.5 M ha is under rubber (*Hevea brasiliensis* (Willd. ex A. Juss.) Müll. Arg.) plantations in Southeast Asia (Zeigler et al., 2009; Tan et al., 2011; De Blecourt et al., 2013). In India, 0.80 M ha of land is currently under rubber plantations. This area has been primarily converted from forest land (The Rubber Board, 2013). Conversion of forests to rubber plantations has been associated with biodiversity loss (Ahrends et al., 2015) and SOC losses of 20–40% (De Blecourt et al., 2013; Guillaume et al., 2015). In contrast, the establishment of rubber plantations on degraded lands has been shown to provide high carbon sink potential (Brahma et al., 2016, 2017). Numerous studies across the world have also indicated increases in SOC stock with an increase in the age of rubber plantations (Zhang et al., 2007; Wauters et al., 2008; De Blecourt et al., 2014).

Quantifying SOC dynamics is important in understanding changes in soil properties and C fluxes in ecosystems. Depending on the residence time of SOC, the C fractions have been categorized as active and passive pools (Chan et al., 2001). The microbial and derivative products, which are very active in recycling within a period of 5 years, have been recognized as the active C (AC) pool, while fractions with relatively longer residence time than the AC pool are characterized as recalcitrant or passive C (PC) pool. These major C pools have differential contributions to atmospheric C concentration due to their release and sequestration processes. The AC pool is the chief source of soil nutrients (Mandal et al., 2008), and therefore it greatly influences the productivity and quality of soil (Mandal et al., 2008; Chan et al., 2001). However, due to its low residence time, the C sink potential of the AC pool is very low (Luo et al., 2003). In addition to improving soil productivity and quality, the PC pool contributes to the total organic carbon (TOC) stock (Mandal et al., 2008). Therefore, the PC pool can act as a reliable indicator of C sequestration potential of a system (Paul et al., 2001). However, information is scanty on the storage of SOC in AC and PC pools under chronosequences of rubber plantation developed on degraded forest land. Therefore, the first objective of this study is to determine the impact of LUC on storage of SOC in the AC and PC pools. Litter production in aboveground (leaf, twig, and branch) and belowground (fine root) plant component depends on land use types, vegetation structure, and composition (Freschet et al., 2013). The availability of organic matter and its decay process determines the quality and quantity of SOC (Krull et al., 2003). For example, high lignin content in belowground litter slows down the decomposition process and subsequently regulates carbon in AC pool (Freschet et al., 2013). The litter decomposition rate is linearly related to turnover rate and inversely related to the mean residence time (MRT). The MRT is influenced by the quality and quantity of litter produced (Liu et al., 2009), which further affects SOC lability (Certini et al., 2015). Therefore, correlating MRT of aboveground and belowground litter with AC pool will reveal the role of litter decomposition in SOC lability. Therefore, the second objective of the present study is to assess the role of aboveground and belowground litter production and its decomposition in the lability of SOC pools.

4.2 METHODS FOR DATA COLLECTION

4.2.1 Study Site

The study area is located in the Barak Valley of Assam in northeast India. It falls within the range of the Himalayan foothills and the Barak River

Basin. The Barak Valley consists of three districts of Assam (Cachar, Hailakandi, and Karimganj) covering an area of ~7000 km^2 (NEDFI, 2016). The valley is characterized by tropical humid climate, and experiences mean annual precipitation of 3500 mm, temperature range of 13–37°C and relative humidity of 93.5% (NEDFI, 2016). The soil texture of the study area is silty clay loam or clay loam. The pH of soil ranges from 4.5 to 5.3. Total nitrogen ranges from 0.45 to 0.78 M kg^{-1} (Vadivelu et al., 2004). The dominant soil is classified as the Barak series, which is fine, mixed, hyperthermic family of *Aeric Endoaquepts* (Soil Survey Staff, 2014), which is the equivalent of Inceptisols in the USDA classification. According to the World Reference Base for Soil Resources (WRB) classification and correlation system (IUSS, 2013), the soils correlate with Cambisols.

The natural forest, adjacent grasslands, and rubber plantations in Karimganj district (24°40′ N, 092°46′ E) were selected for the present study. The forest is classified as tropical wet evergreen forest (Champion and Seth, 2005), with the dominant tree species being *Artocarpus chama* (Buch.-Ham. ex Wall.), *Cynometra polyandra* (Roxb. (Harms)), *Ficus hispida* (Linn. f.), *Lindera sp.*, *Litsea sp.*, *Oroxylum indicum* ((L.) Kurz)), and *Syzygium cumini* ((L.) Skeels) and so on. The grasslands in Barak Valley originally developed as a result of degradation of natural forests and subsequently invaded by *Imperata. cylindrica*. Annual harvesting of aboveground parts of the *I. cylindrica* for thatching and roofing purposes is very common in North Eastern India (Pathak et al., 2017). Burning of aboveground residues is also a common managerial system practiced annually to maintain the soil nutrients under *I. cylindrica* grasslands (Pathak et al., 2017). The rubber plantations were developed on grassland that developed following degradation of the natural forest. Rubber plantations were established for the production of natural rubber latex, which is an important source of monitory income in the region. For the present study, rubber plantations of different ages situated at 200–300 m from each other were chosen. The age of the rubber tree plantations and the history of the land use were confirmed from the forest's official records and interview of local people.

4.2.2 Soil Sampling and Analyses

Soil samples were collected from each of the study sites (rubber plantation, natural forest, and grassland) during the months of February and March in

2016. Four 250 m × 250 m quadrants were selected from each plantation and land uses following the ISRO-GBP/NCP-VCP protocol described in Singh and Dadhwal (2009). The quadrants were chosen on representative sites covering dense to sparse vegetation cover. Since the selected rubber plantations were situated on slopes of small hillocks, the quadrants were laid in such a way to cover the top, middle, and base of the hillock. The distance between any two representative quadrants within each plantation ranged between 30 and 50 m. Within each quadrant three vertical soil profiles (1 m × 1 m ×1 m) were dug at the top, middle, and base of the slope. Thus, for each rubber plantations, a total of 12 vertical profiles (four quadrants × three soil profiles) were dug. The natural forest and *I. cylindrica* grasslands were situated on plain landscapes, therefore, two profiles were dug within each of the 250 m × 250 m quadrants. Thus, in case of the natural forest and *I. cylindrica* grassland, eight vertical profiles (four quadrants × two soil profiles) were dug for collecting soil samples.

Soil samples were collected using 10 cm scaled soil cores with 5.6 cm inner diameter from the 0–20 cm, 20–50 cm, and 50–100 cm depth of each soil profile. Three samples were collected by inserting the core horizontally up to 10 cm and composite soil samples were prepared for each of the soil depth. In total, 36 composite samples were collected for each of the rubber plantations. In case of natural forest and *I. cylindrica* grassland, 24 composite samples each were collected and taken to the laboratory for analysis. The soil samples were air-dried and sieved through 150 μm (100 mesh) sieve for determination of SOC stock and C pools. Coarse mineral fragments (>2 mm) were excluded from SOC stock calculation.

SOC concentration of each of the soil samples was determined through the wet oxidation method (Walkley and Black, 1934; Nelson and Sommers, 1982). A modified Walkley and Black method as described in Chan et al. (2001) was used to determine the proportion of AC and PC pools. Different acid-aqueous solution results of 12, 18, 24, and 36 N of H_2SO_4 were applied, and according to their decreasing order of oxidation ability, four different pools were extracted. Organic C oxidized by 12N H_2SO_4 has been considered as a very labile C (VLC) fraction, whereas the difference between C oxidized by 18 and 12 N has been considered as labile C (LC). Cumulative values for VLC and LC has been considered as AP. In addition, the difference between C oxidizable by 24 and 18 N has been considered as less-labile C (LLC) and the difference between 36 and

24 N oxidized C has been considered as a non-labile C (NLC) fraction. The sum of LLC and NLC is considered as PC pool (Chan et al., 2001).

The SOC stocks (Mg ha^{-1}) for each depth under different land uses were computed following the method of Blanco-Canqui and Lal (2008) as follows:

$$\text{SOC stock} \left(\text{Mg ha}^{-1} \right) = 10^4 \left(\frac{m^2}{ha} \right) \times \text{soil depth (m)} \times \text{BD} \left(\frac{Mg}{m^3} \right) \times \frac{\text{SOC}(\%)}{100}$$

where BD is bulk density and SOC is soil organic carbon concentration.

SOC stocks for 0–100 cm soil depths for each land uses were estimated by summing up the SOC stocks of the different depths (0–20, 20–50, and 50–100 cm). For estimating soil bulk density, two separate samples were collected from each depth using the same soil core. A total of 72 soil samples (12 profiles × three depths × two samples from each depth) were collected from each of the rubber plantation, whereas a total of 48 soil samples (eight profiles × three depths × two samples from each depth) were collected from the natural forest and *I. cylindrica* grassland separately for BD analysis. In the laboratory, soil BD was estimated using the method described by Robertson et al. (1974).

4.2.3 Aboveground Litter Production

Five 1 m × 1 m quadrants were laid in each quadrant measuring 250 m × 250 m for different aged rubber plantations. Thus, from each rubber plantation, aboveground litter (i.e., dead leaves and branches) was collected from a total of 20 quadrants at 30 days interval for 1 year. The time period was extended up to maximum of 32 days when the field visit was not possible at the scheduled interval due to rain. Litter samples were taken to the laboratory to obtain the dry weight after drying in a closed oven at 70°C for 72 h. Litter collected from all the quadrants were averaged and scaled up to Mg ha^{-1}year^{-1}.

4.2.4 Belowground Litter Production

Fine roots of the surface soil layer are very dynamic in nature and play a significant role in soil C flux (Chimento and Amaducci, 2015). Therefore, dead fine root production of rubber plantations was measured in the top 0–20 cm soil depth for estimation of belowground litter. Soil samples were collected from 0 to 20 cm depth of the soil with a sharp core of 5.6 cm inner diameter to extract the fine roots. For belowground litter collection, three plots with 10 m × 10 m area were randomly laid within each of the 250 m × 250 m quadrants. During each of the monthly sampling dates, two random samples were collected from each of the 12 10 m × 10 m plots. Therefore, a total of 24 core samples were collected at monthly interval from each rubber plantation and taken to the laboratory for further analysis. In the laboratory, soil samples were soaked overnight and roots were cleaned from soil and organic residues by a gentle stream of water over a sieve. Roots with ≤2 mm diameter were separated manually. Live and dead roots were also carefully separated soon after removing the roots of other plant material. Live roots of rubber plants were directly observed as light in color, resilient to breakage and decay free, whereas the dead roots were brown or black in color and easily broken (Bennett et al., 2002; Brassard et al., 2011). Fine roots of other herbs and shrubs were separated based on color, branching pattern, and texture. For estimating dead fine root biomass, dry weight of the fine roots was taken separately after drying at 70°C for 72 h (Jamro et al., 2015) and averaged for reporting as monthly biomass. Annual dead fine root biomass (RB) was estimated by subtracting the minimum biomass (RB_{min}) from the maximum biomass (RB_{max}) recorded over the course of the sampling period (McClaugherty et al., 1982; Publicover and Vogt, 1993; Lehmann and Zech, 1998). Aboveground and belowground litter biomass production for natural forest and *I. cylindrica* grassland sites was not estimated since the focus of this study was to explore the role of aboveground and belowground litter production and its decomposition in the lability of SOC pools under chronosequence of rubber plantation.

4.2.5 MRT Determination

Aboveground and belowground litter decomposition under different age of rubber plantation has been determined through mesh bag technique (Fahey et al., 1988). Around 150 numbers of mesh bags for each category of litter were prepared (Jamro et al., 2015). Each of the aboveground litter bags contained 4 g of aboveground litter biomass and belowground litter bag contained 2 g of belowground litter biomass. The litter bags were then equally distributed and placed in three different places under each rubber plantation. At the monthly sampling date, three litter bags for each category were collected from all the three different places. Therefore, nine mesh bags were collected at monthly interval for aboveground and belowground each, from a rubber plantation. At each sampling date, a total of 72 numbers of mesh bags (36 for each category) were collectively retrieved from all the rubber plantations. During the laboratory analysis, residual litters were carefully removed from the bag and washed gently over the mesh to remove the adherent soils. Residual roots were dried at 70°C for 72 h to obtain the dry weight of each sample.

MRT of aboveground and belowground litters under different ages of rubber tree plantations was calculated using the following formula (Giardina and Ryan, 2000):

$$MRT = 1/k,$$

where k is the litter decomposition rate constant, which was estimated using the following formula (Wieder and Lang, 1982):

$$M_t/M_0 = e^{-k,t},$$

where M_0 is the initial dry mass of the fine root, M_t is the remaining fine root dry mass at t time.

4.2.6 Statistical Analyses

The data collected were subjected to two-way analysis of variance using the generalized linear model (GLM procedure) to determine the effect of chronosequence and soil depth on the different C pools and SOC stocks. When the F ratio indicated significance at $P = 0.05$, means were separated using Tukey's honestly significant difference (HSD) test. Tukey's method was used in preference to other methods because it is now the most acceptable method for all pair-wise comparisons. Unlike most methods, this method considers the statistical distributions associated with repeated testing and it provides an exact adjusted P values (Hsu, 1996). For comparing means in figures the 95% confidence intervals (CI) were used preference to other methods because 95% CI functions as a very conservative test of hypothesis and it also attaches a measure of uncertainty to sample statistic (du Prel et al., 2009). Two means were considered significantly different from each other only when their 95% CI does not overlap.

Correlation analysis was applied to explore the relationship between AC pools in the upper soil layer (0–50 cm) with aboveground and belowground litter production and MRT.

Since the rubber trees are being managed in a plantation up to a production age of 35–40 years (Brahma et al., 2016), we estimated the time-averaged SOC sequestration rate for the 34-year-old plantations. SOC stocks under 27-year-old plantations were subtracted from the SOC stocks of 34-year-old plantation and the result was averaged by the number of years from 27 to 34 years, which was considered as SOC sequestration rate of 34-year-old rubber tree plantation. Finally, to predict the time required to achieve equilibrium, the differences in SOC stock between 34-year-old rubber plantation and that under a forest was divided by the SOC sequestration rate of 34-year-old rubber plantation.

4.3 VARIATION IN C POOLS

The different fractions of soil C significantly ($P < 0.001$) varied with chronosequence and soil depth (Table 4.1; Fig. 4.1). In order of increasing lability, different SOC fractions in the different land used for 0–100 cm soil depth are shown in Figure 4.1a–d. The highest values of VLC (31.7 Mg ha^{-1}) and LC (30.8 Mg ha^{-1}) were recorded under 6-year-old plantations. On the other hand, the highest values of LLC (33.2 Mg ha^{-1}) and NLC (53.8 Mg ha^{-1}) were recorded under 34-year-old plantations (Fig. 4.1). SOC stocks under the rubber plantations varied from 106.2 Mg ha^{-1} in 6-year-old plantations to 130.3 Mg ha^{-1} in 34-year-old plantations in the top 100 cm soil depth (Fig. 4.1e). The SOC stocks under natural forest and *I. cylindrica* grassland were estimated at 197.6 and 108.2 Mg ha^{-1}, respectively (Fig. 4.1e). The proportion of NLC to the SOC stocks varied between 29% and 37% in the 0–20 cm, 14% and 23% in the 20–50 cm, and 21% and 57% in the 50–100 cm soil depths (Fig. 4.2). Similarly, the proportion of LLC to the SOC stocks varied between 12% and 15% in the 0–20 cm, 22% and 34% in the 20–50 cm and 23% and 30% in the 50–100 cm soil depths (Fig. 4.2). The VLC and LC fractions in the 0–100 cm depth decreased from 30% to 13% and 29% to 21%, respectively under

Impact of Land Use Changes

6- and 34-year-old rubber plantations. On the other hand, the LLC and NLC fractions increased from 20% to 25% and from 21% to 41% under 6- and 34-year-old rubber plantations, respectively (Fig. 4.2).

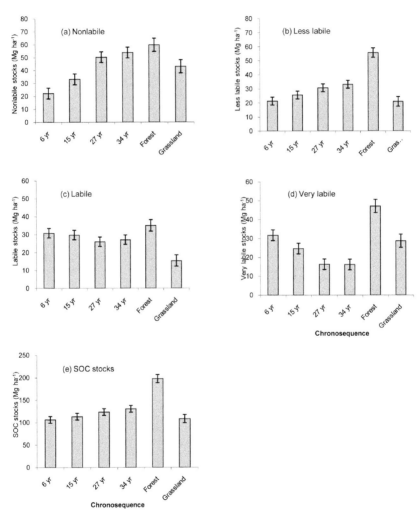

FIGURE 4.1 Variations in stocks (Mg ha^{-1}) of very labile carbon (VLC), labile carbon (LC), less labile carbon (LLC), non-labile carbon (NLC), and SOC stocks with chronosequence in the 0–100 cm depth. Error bars represent 95% confidence limits. Two means are considered significantly different from each other only when their error bars do not overlap.

Source: Reprinted with permission from Nath et al., 2017. © 2017 Elsevier B.V.

TABLE 4.1 Variations in Stocks (Mg ha^{-1}) of Very Labile Carbon (VLC), Labile Carbon (LC), Less Labile Carbon (LLC), and Nonlabile Carbon (NLC) with Chronosequence and Soil Depths.

Soil Depth	Chronosequence	Non-labile C	Less labile C	Labile C	Very labile C
0–20	6 years	7.4 (0.37) c	3.1 (0.33) c	10.2 (0.42) b	5.0 (0.28) c
	15 years	10.4 (0.68) b	5.2 (0.30)bc	10.4 (0.48) ab	7.5 (0.49)bc
	27 years	14.9 (0.38) a	6.0 (0.48) ab	10.8 (0.65) ab	9.0 (0.86) b
	34 years	15.7 (0.94) a	6.6 (0.87) ab	12.9 (0.93) a	9.7 (0.86) ab
	Forest	10.8 (0.99) b	8.2 (0.50) a	7.5 (0.23) c	13.5 (1.71) a
	Grassland	16.9 (0.96) b	6.0 (0.53) ab	5.5 (0.77) c	10.6 (1.22) ab
	F value	26.0	8.9	15.2	9.7
	P value	<0.001	<0.001	<0.001	<0.001
20–50	6 years	4.0 (0.26) d	6.5 (0.34) c	9.3 (0.44) a	9.5 (0.47) bc
	15 years	4.3 (0.68) d	8.4b (0.55)c	10.0 (0.57) a	7.2 (0.47) c
	27 years	6.4 (0.46) cd	10.4 (1.20) b	11.1 (1.83) a	3.2 (0.47) d
	34 years	8.0 (1.36) bc	11.3 (0.69) b	10.5 (0.76) a	3.3 (0.33) d
	Forest	21.1 (1.33) a	17.1 (0.91) a	2.1 (0.39) b	21.2 (0.87) a
	Grassland	11.3 (0.64) b	5.5 (0.67) c	3.6 (0.44) b	10.1 (1.27) b
	F value	46.2	23.0	13.3	98.1
	P value	<0.001	<0.001	<0.001	<0.001
50–100	6 years	10.8 (0.98) c	11.8 (0.44)bc	11.4 (0.91) b	17.1 (1.35) a
	15 years	18.5 (1.48) b	12.0 (1.06)bc	9.1 (0.98) bc	9.9 (0.75)bc
	27 years	29.0 (1.28) a	14.4 (0.55) b	4.2 (0.56) d	4.1 (0.41) d
	34 years	30.1 (2.81) a	15.3 (1.12) b	3.7 (0.60) d	3.1 (0.49)d
	Forest	27.8 (1.16) a	30.6 (1.81) a	25.9 (2.22) a	12.4 (1.06) b
	Grassland	14.8 (1.02)bc	9.6 (1.03) c	6.3 (0.65) cd	8.0 (0.90) c
	F value	23.7	45.0	53.0	39.1
	P value	<0.001	<0.001	<0.001	<0.001

Figures followed by the same letters in a column in each depth class are not significantly different from each other according to Tukey's HSD. Figures in parenthesis represent standard errors of means.

Source: Reprinted with permission from Nath et al., 2017. © 2017 Elsevier B.V.

Impact of Land Use Changes

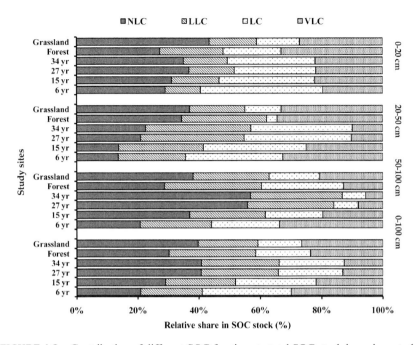

FIGURE 4.2 Contribution of different SOC fractions to total SOC stock in various study sites. NLC, non-labile carbon; LLC, less labile carbon; LC, labile carbon; and VLC, very labile carbon.

Source: Reprinted with permission from Nath et al., 2017. © 2017 Elsevier B.V.)

The AC and PC pools also significantly varied with the chronosequence and soil depth (Fig. 4.3). In natural forest, both AC and PC pools increased with soil depth, but for *I. cylindrica* grassland the opposite trend was followed for the upper two depths (Fig. 4.3). AC pool in 0–20 cm soil depth was significantly increased from 15.16 Mg ha^{-1} under 6 years to 22.62 Mg ha^{-1} under 34-year-old plantation (Fig. 4.3a), whereas, for 20–50 cm and 50–100 cm soil depth the values were significantly decreased with the plantation age (Fig. 4.3a). AC pool in 0–20 cm of natural forest was not significantly different from 34-year-old plantation, but in 20–50 cm and 50–100 cm soil depth the values were significantly higher than that of any other land use of the present study (Fig. 4.3a). The PC pools under rubber plantations significantly increased with plantation age in the same soil depth (Fig. 4.3b). However, in 50–100 cm soil depth, the PC pools under 27–34-year-old plantations did not significantly differ from each other. The PC pool in 0–20 cm soil depth under the natural forest was not significantly different from 27 years, 34 years, and grassland. On the other hand, the PC pools in 20–50 cm and 50–100 cm soil depths under natural forest were significantly higher than that of under all other land uses (Fig. 4.3b). Over 59% of the total SOC stock in the 0–100 cm depth was contributed by the AC pools under 6-year-old plantation that decreased to 33% under 34-year-old plantation (Fig. 4.4). Allocation of AC pools in the same soil depth under natural forest and *I. cylindrica* grassland were estimated at 42% and 41% of their respective SOC stocks (Fig. 4.4). In addition to that, the contribution of VLC to AC declined from 51% to 37% in 6- to 34-year-old plantations, respectively. Results further showed the contribution of NLC in PC increased from 51% to 62% from 6 to 34year-old plantations (Table 4.1; Fig. 4.4).

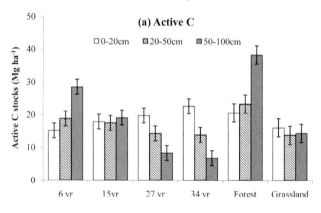

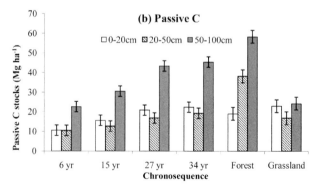

FIGURE 4.3 Variations in passive and active C stocks (Mg ha^{-1}) with chronosequence and soil depth. Error bars represent 95% confidence limits. Two means are considered significantly different from each other only when their error bars do not overlap.

Source: Reprinted with permission from Nath et al., 2017. © 2017 Elsevier B.V.

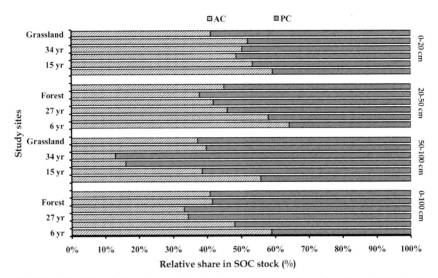

FIGURE 4.4 The relative shares of active C pool (AC) and passive C pool (PC) with respect to soil depth under the different ages of rubber tree plantations, natural forest, and grasslands of northeast India.

Source: Reprinted with permission from Nath et al., 2017. © 2017 Elsevier B.V.

4.4 LITTER PRODUCTION

Aboveground and belowground litter production increased with increase in plantation age. In contrast, proportion of AC pool decreased with increase in plantation age (Fig. 4.5). Aboveground litter biomass under 6–34-year-old rubber plantations ranged between 7.95 (± 0.51) and 10.89 (± 0.15) Mg ha^{-1}year^{-1} and the respective values for belowground litter biomass ranged between 0.55 (± 0.04) and 0.95 (± 0.09) Mg ha^{-1}year^{-1} (Table 4.2). However, litter biomass productions were negatively correlated ($P < 0.05$) with the AC pool in the different aged rubber plantations (Table 4.2). The correlation between AC pools on one hand and aboveground and belowground litter MRT was positive ($r = 0.988$ and 0.915, respectively, $P < 0.05$).

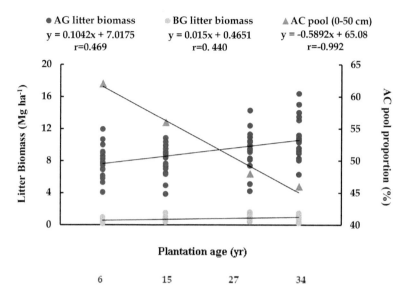

FIGURE 4.5 Trend showing correlations of aboveground litter production, belowground litter production, and active carbon pool with different aged rubber plantation.

TABLE 4.2 Aboveground and Belowground Litter Production and Mean Residence Time (MRT) Under Rubber Plantations.

Stand age (years)	Litter production (Mg ha^{-1} year^{-1})		Mean residence time of litter (years)	
	Aboveground	Belowground	Aboveground	Belowground
6	7.95 (0.41)b	0.55 (0.07)a	1.01	0.96
15	8.24 (0.42)b	0.41 (0.02)a	0.68	0.82
27	9.53 (0.52)ab	0.91 (0.11)b	0.44	0.81
34	10.89 (0.55)a	0.95 (0.10)b	0.38	0.73
F value	7.9	10.42	NA	NA
P values	<0.0001	<0.0001	NA	NA
Correlation with AC	−0.924	−0.997	0.988	0.915

Source: Reprinted with permission from Nath et al., 2017. © 2017 Elsevier B.V.

Figures followed by the same letter in a column are not significantly different from each other according to Tukey's HSD test. Figures in parenthesis represent standard errors of means. The correlation coefficients (r)

of AC pools with aboveground and belowground litter production and MRT of litter in the 0–50 cm soil depth are given in the lower part of the table. NA = not applicable.

4.5 SOC LOSSES AND SEQUESTRATION POTENTIALS

Analysis of changes in SOC stock due to LUC from natural forest to rubber plantations revealed a potential net loss of 67.3 Mg C ha^{-1} at the age of 34-year-old plantation (Fig. 4.1a). A cumulative loss of 89.5 Mg C ha^{-1} from all the soil depths were observed when natural forest was degraded and subsequently invaded by *I. cylindrica* grassland (Fig. 4.1a). The conversion of natural forest into rubber plantations resulted in net loss of 91.5 Mg ha^{-1} of SOC after 6 years of establishment and it decreased to the 67.40 Mg ha^{-1} after 34 years of plantation establishment.

The potential SOC sequestration rate varied with plantation age and it ranged between 0.18 and 0.99 Mg ha^{-1} year^{-1} for 6 and 34-year-old plantations, respectively. Based on the C sequestration rate of 0.99 Mg ha^{-1} year^{-1} under 34-year-old rubber plantations, it is predicted that rubber plantations can achieve the equilibrium in SOC stock with that under natural forest if they can be managed for additional 68 years after its productive age.

4.6 CHANGES IN SOC STOCK AND LABILITY

The analysis revealed significant variations in SOC stocks with chronosequence. As expected, natural forest accumulated the highest SOC stocks than rubber plantations and *I. cylindrica* grassland, particularly in the 50–100 cm depth. The SOC stock of 198 Mg ha^{-1} recorded under natural forest in the present study is comparable with other forest sites in India (Saha et al., 2011). The estimated SOC stocks under rubber plantations from different parts of the world varied between 135 and 194 Mg ha^{-1} under 5- and 44-year-old plantations (Wauters et al., 2008; Egbel et al., 2012; De Blecourt et al., 2013; Chiti et al., 2014). The conversion of forests into rubber plantation may result 15% to 50% SOC loss over the period (De Blecourt et al., 2013; Guillaume et al., 2015; Li et al., 2015; Straaten et al., 2015). Loss of SOC from the 0–50 cm soil depth was higher than that of the lower soil depths. The loss of SOC from the upper soil depth might have been triggered by loss of vegetative cover due to clearing of vegetation prior to the rubber plantation development (Brahma et al., 2016) or removal of aboveground parts of *I. cylindrica* grasses resulting from annual burning management system (Pathak et al., 2017). The removal of vegetative cover may cause SOC loss as high as 50% when natural forest is converted to other land uses (Straaten et al., 2015). Similar studies from India have reported loss of 23% of SOC stock when natural forests were converted into grasslands (Iqbal and Tiwari, 2016). Relatively high SOC losses from the subsurface are due to lower rate of production of fine roots and subsequently less SOM in such layer because of their shallow root nature (Pathak et al., 2017).

Development of plantations in grassland may increase the C sink in the soil through increase of biomass (Lal, 2005). In the present study, the SOC stock recorded in 6-year-old rubber plantation was 2% lower than that under *I. cylindrica* grassland. This may be due to loss of C from the upper layer through removal of aboveground vegetation, intensive sunlight exposures to the field, and purposive soil disturbances during the plantation development period. A comparable result was observed with 25% higher SOC stock under 15-year-old rubber plantation than the adjacent pasture land in Brazil (Maggiotto et al., 2014). Moreover, lower canopy density in younger stands of rubber plantation (Li and Fox, 2012) lead the surface soil exposed to the sunlight, resulted in loss of SOC from the upper layer (0–20 cm) even after 15 years of age.

LUC may affect the systems management practices, vegetative characteristics, nutrient status, and SOC stock (Chen et al., 2016), which may collectively reflect on the allocation of different SOC fractions (Sarkar et al., 2015). However, the relative proportion of AC pools (VLC + LC) and PC pools (LLC + NLC) depends on the availability of nutrients in the soil (Lian and Zhang, 1998). The efficiency of micro-organisms may be impaired within nutrient-rich soils that may lead to lowering the proportion of SOC accumulation in AC pools (Chen et al., 2016). The soils under natural forests are known to be nutrient-rich through the inputs from aboveground and belowground litter (Lian and Zhang, 1998). In the present study, 42% of the total SOC stock was found to be allocated in AC pools under natural forest, which was close to that under *I. cylindrica* grassland, but lower than that of 6- and 15-year-old rubber plantations. The similarity in the proportion of AC pool in natural forest and *I. cylindrica* grassland is probably due to annual burning of aboveground biomass in the latter (Pathak et al., 2017), which supplement particulate organic matter and other easily metabolized organic compounds such as sugar and amino acids into the soils and subsequently enhances AC pool. Higher AC pool under younger rubber stands than the natural forest and *I. cylindrica* grassland is probably due to the efficient use of organic carbon stocks in the former in order to fulfill the systems nutrient requirements by the heterotrophs (Chen et al., 2016). The present study revealed that 41–67% of SOC was stored in recalcitrant or PC pool under rubber plantations. The leaves of rubber trees have been reported to contain tannin (2–4% on dry matter base) (Abulude, 2007; Wigati et al. 2014). A positive priming effect of tannin and waxy compounds (i.e., lipids) in leaf litter may increase the SOC content of the soil in PC or recalcitrant pool (Braids and Miller, 1975; Almendros et al., 2001; Gleixner et al., 2001; Kraus et al., 2003). The proportion of high SOC in PC pools is attributed to the chemical composition of rubber leaf with high tannin and wax content. The share of NLC and LLC to the SOC stocks under rubber plantations increased with increase in the age of the plantation and soil depth. Since, PC pools are less oxidative than AC pools (Sarkar et al., 2015), the high proportion of PC pools may act as an indicator of the relative stability of the SOC stocks in a system. The high proportion of NLC and LLC in the 0–100 cm soil depth under 27- and 34-year-old stands is probably indicative of achievement of equilibrium state under this land use. The higher C fractions in PC pools under mature rubber plantations are attributed to the deeper root penetrations. Moreover, low decomposition rate of roots due to lower activity of microorganisms and lower oxygen level may constrain the loss of soil C from the lower depth (Spohn et al., 2016). Therefore, the observed increase in PC pools after rubber plantation development from the forests and grasslands signifies the development of a SOC stable system over the year.

Aboveground and belowground litter is the most important sources of SOC (Wang et al., 2013) and its production is induced by the vegetative characteristics of the stand (Wang and Wang, 2007). Litter quantity and quality have significant influence on the soil microbial biomass and its activity, which further influence the SOC and its stability in the soil (Vesterdal et al., 2008; Wang et al., 2013). Scarcity of litter inputs may force the lability of the soil C pools and trigger C losses from the soil (Chan et al., 2001). The upper layer of the soil C is mostly influenced by the aboveground and belowground litter productions and its subsequent decomposition, and the soil C fractions in deeper soil depths are mostly influenced by the previous land uses (Fisher et al., 1994). Results revealed the aboveground and belowground litter biomass production increased with the increase in age of the rubber plantation. Comparatively low litter production in the younger stands triggered production of more C under AC pool to fulfill nutrient requirement of the growing trees. The negative correlationship between AC pool and litter production further strengthen the argument that low litter production conserves more SOC under AC pool and vice versa.

4.7 CONCLUSIONS

Based on the analysis the following conclusions were drawn: (1) rubber plantations achieve higher SOC stocks than *I. cylindrica* grassland but lower SOC stocks than natural forests, (2) the proportion of recalcitrant C pools increase with increase in age of the rubber plantation, and (3) a negative correlation between AC pool and total aboveground and belowground litter production suggests low litter production at the initial plantation age conserves more SOC in AC pool.

KEYWORDS

- active carbon
- recalcitrant carbon
- residence time
- litter production
- lability index

REFERENCES

Abera, G.; Wolde-Meskel, E. Soil Properties, and Soil Organic Carbon Stocks of Tropical Andosol Under Different Land Uses. *Open J. Soil Sci.* **2013**, *3*, 153–162.

Abulude, F. O. Phytochemical Screening and Mineral Contents of Leaves of Some Nigerian Woody Plants. *Res. J. Phytochem.* **2007**, *1*, 33–39.

Ahrends, A.; Hollingsworth, P. M.; Ziegler, A. D.; Fox, J. M.; Chen, H.; Su, Y.; Xu, J. Current Trends of Rubber Plantation Expansions May Threaten Biodiversity and Livelihoods. *Glob. Environ. Chang.* **2015**, *34*, 48–58.

Almendros, G.; Tinoco, P.; Gonzalez-Vila, F. J.; Ludemann, H. D.; Sanz, J.; Velasco, F. 13C-NMR of Forest Soil Lipids. *Soil Sci. Soc. Am. J.* **2001**, *166*, 186–96.

Bennett, J. N.; Andrew, B.; Prescott, C. E. Vertical Fine Root Distributions of Western Redcedar, Western Hemlock, and Salal in Old-growth Cedar-hemlock Forests on Northern Vancouver Island. *Can. J. For. Res.* **2002**, *32*, 1208–1216.

Blanco-Canqui, H.; Lal, R. No-tillage and Soil-profile Carbon Sequestration: An On-farm Assessment. *Soil Sci. Soc. Am. J.* **2008**, *72*, 693–701.

Brahma, B.; Nath, A. J.; Das, A. K. Managing Rubber Plantations for Advancing Climate Change Mitigation Strategy. *Curr. Sci.* **2016**, *110*, 2015–2019.

Brahma, B.; Sileshi, G. W.; Nath, A. J.; Das, A. K. Development and Evaluation of Robust Tree Biomass Equations for Rubber Tree (*Hevea brasiliensis*) Plantations in India. *Forest Ecosyst.* **2017**.

Braids, O. C.; Miller, R. H. Fats, Waxes, and Resins in Soil. In *Soil Components*; Gieseking, J. E., Ed.; Springer: Berlin, 1975; pp 343–68.

Brassard, B. W.; Chen, H. Y.; Bergeron, Y.; Pare, D. Differences in Fine Root Productivity Between Mixed and Single Species Stands. *Func. Ecol.* **2011**, *25*, 238–246.

Certini, G.; Vestgarden, L. S.; Forte, C.; Strand, L. T. Litter Decomposition Rate and Soil Organic Matter Quality in a Patchwork Healthland of Southern Norway. *Soil* **2015**, *1*, 207–216.

Champion, H. G.; Seth, S. K. *A Revised Survey of the Forest Types of India.*; Natraj Publishers: Dehradun, 1968; p 404 (Reprinted 2005).

Chan, K. Y.; Bowman, A.; Oates, A. Oxidizible Organic Carbon Fractions and Soil Quality Changes in an Oxic Paleustalf Under Different Pasture Leys. *Soil Sci.* **2001**, *166*, 61–67.

Chen, G.; Yang, Y.; Yang, Z., Xie, J.; Guo, J.; Gao, R.; Yin, Y.; Robinson, D. Accelerated Soil Carbon Turnover Under Tree Plantations Limits Soil Carbon Storage. *Sci. Rep.* **2016**, *6*, 19693.

Chimento, C.; Amaducci, S. Characterization of Fine Root System and Potential Contribution to Soil Organic Carbon of Six Perennial Bioenergy Crops. *Biomass Bioenergy* **2015**, *83*, 116–122.

Chiti, T.; Grieco, E.; Perugini, L.; Rey, A.; Valentini, R. Effect of the Replacement of Tropical Forests with Tree Plantations on Soil Organic Carbon Levels in the Jomoro District, Ghana. *Plant Soil* **2014**, *375*, 47–59.

De Blécourt, M.; Brumme, R.; Xu, J.; Corre, M. D.; Veldkamp, E. Soil Carbon Stocks Decrease Following Conversion of Secondary Forests to Rubber (*Hevea brasiliensis*) Plantations. *PLoS One* **2013**, *8*, e69357.

De Blecourt, M.; Hansel, V. M.; Brumme, R.; Corre, M. D.; Veldkamp, E. Soil Redistribution by Terracing Alleviates Soil Organic Carbon Losses Caused by Forest Conversion to Rubber Plantation. *For. Ecol. Manage.* **2014**, *313*, 26–33.

duPrel, J. B.; Hommel, G.; Röhrig, B.;Blettner, M. Confidence interval or P-value? *Deutsc. Ärzt. Int.* 2009, 106, 335–9.

Egbel, A. E.; Tabot, P. T.; Fonge, B. A.; Becham, E. Simulation of the Impacts of Three Management Regimes on Carbon Sinks in Rubber and Oil Palm Plantation Ecosystems of Southwestern Cameroon. *J. Ecol. Nat. Environ.* **2012**, *46*, 154–162.

Fahey, T. J.; Hughes, J. W.; Mou, P.; Arthur, M. A. Root Decomposition and Nutrient Flux Following Whole-Tree Harvest of Northern Hardwood Forest. *For. Sci.* **1998**, *34*, 744–768.

Fan, S.; Guan, F.; Xu, X.; Forrester, D. I.; Ma, W.; Tang, X. Ecosystem Carbon Stock Loss After Land Use Change in Subtropical Forests in China. *Forests* **2016**, *7*, 142.

Fisher, M. J.; Rao, I. M.; Ayarza, M. A.; Lascano, C. E.; Sanz, J. I.; Thomas, R. J.; Vera, R. R. Carbon Storage by Introduced Deep-rooted Grasses in South African Savannas. *Nature* **1994**, *371*, 236–238.

Freschet, G. T.; Cornwell, W. K.; Wardle, D. A.; Elumeeva, T. G.; Liu, W.; Jackson, B. G.; Onipchenko, V. G.; Soudzilovskaia, N. A.; Tao, J.; Cornelissen, J. H. C. Linking Litter Decomposition of Above- and Below-Ground Organs to Plant–Soil Feedbacks Worldwide. *J. Ecol.* **2013**, *101*, 943–952.

Giardina, C. P.; Ryan, M. G. Evidence that Decomposition Rates of Organic Carbon in Mineral Soil Do not vary with Temperature. *Nature* **2000**, *404*, 858–861.

Gleixner, G.; Czimczik, C. J.; Kramer, C.; Luhker, B.; Schmidt, M. W. I. Plant Compounds and their Turnover and Stabilization as Soil Organic Matter. In *Global Biogeochemical Cycles in the Climate System*; Schulze, M. H. E. D., Harrison, S., Holland, E., Lloyd, J., Prentice, I. C., Schimel, D., Eds.; Academic Press: San Diego, 2001; pp 201–215.

Guillaume, T.; Muhammad, D.; Kuzyakov, Y. Losses of Soil Carbon by Converting Tropical Forest to Plantations: Erosion and Decomposition Estimated by $\Delta 13C$. *Glob. Chang. Biol.* **2015**.

Hsu, J. C. *Multiple Comparisons*; Chapman and Hall: London,1996.

Iqbal, S.; Tiwari, S. C. Soil Organic Carbon Pool Under Different Land Uses in Achanakmar Amarkantak Biosphere Reserve Of Chhattisgarh, India. *Curr. Sci.* **2016**, *110*, 771–73.

ITTO. ITTO Guidelines for the Restoration, Management and Rehabilitation of Degraded and Secondary Tropical Forests, Policy Development Series No. 13. Yokohama: ITTO; 2002.

IUSS. *World Reference Base for Soil Resources 2014*. International soil classification system for naming soils and creating legends for soil maps, IUSS Working Group, World Soil Resources Reports No. 106; FAO: Rome, 2014.

Jamro, G. M; Chang, S. X.; Naeth, M. A.; Duan, M.; House, J. Fine Root Dynamics in Lodgepole Pine and White Spruce Stands Along Productivity Gradients in Reclaimed Oil Sands Sites. *Ecol. Evol.* **2015**, *5* (20), 4655–4670.

Jobbágy, E. G.; Jackson, R. B. The Vertical Distribution of Soil Organic Carbon and Its Relation to Climate and Vegetation. *Ecol. Appl.* **2000**, *10*, 423–436.

Kraus, T. E. C.; Dahlgren, R. A.; Zasoski, R. J. Tannins in Nutrient Dynamics of Forest Ecosystems: A Review. *Plant Soil* **2003**, *256*, 41–66.

Krull, E. S.; Baldock, J. A.; Skjemstad, J. O. Importance of Mechanisms and Processes of the Stabilisation of Soil Organic Matter for Modelling Carbon Turnover. *Funct. Plant Biol.* **2003**, 30 (2), 207–222.

Lal, R. Forest Soils and Carbon Sequestration. *For. Ecol. Manage.* **2005**, *220*, 242–258.

Lal, R. Soil Health and Carbon Management. *Food Energy Secur.* **2016**, *5* (4), 212–222. DOI: 10.1002/fes3.96.

Lehmann, J.; Zech, W. Fine Root Turnover of Irrigated Hedgerow Intercropping in Northern Kenya. *Plant Soil* **1998**, *198*, 19–31.

Li, Z.; Fox, J. M. Mapping Rubber Tree Growth in Mainland Southeast Asia Using Time-Series MODIS 250 m NDVI and Statistical Data. *Appl. Geo.* **2012**, *32*, 420–432.

Li, S.; Su, J.; Liu, W.; Lang, X.; Huang, X.; Jia, C.; Zhang, Z.; Tong, Q. Changes in Biomass Carbon and Soil Organic Carbon Stocks Following the Conversion from a Secondary Coniferous Forest to a Pine Plantation. *PLoS One* **2015**, *10*, E0135946.

Lian, Y.; Zhang, Q. Conversion of a Natural Broad-Leafed Evergreen Forest into Pure and Mixed Plantation Forests in a Subtropical Area: Effects on Nutrient Cycling. *Can. J. For. Res.* **1998**, *28* (10), 1518–1529.

Lui, L.; King, J. S.; Giardina, C. P.; Booker, F. L. The Influence of Chemistry, Production and Community Composition Under Elevated Atmospheric CO_2 and Tropospheric O_3 in a Northern Hardwood Ecosystem. *Ecosystems* **2009**, *12*, 401–416.

Luo, Y.; White, L. W.; Canadell, J. G.; DeLucia, E. H.; Ellsworth, D. S.; Finzi, A.; Lichter, J.; Schlesinger, W. H. Sustainability of Terrestrial Carbon Sequestration: A Case Study in, Duke Forest with Inversion Approach. *Glob. Biogeochem. Cycles* **2003**, *17*, 1021.

Maggiotto, S. R.; Oliveira, D.; Marur, C. J.; Stivari, S. M. S.; Leclerc, M.; Wagner-Riddle, C. Potential Carbon Sequestration in Rubber Tree Plantations in the Northwestern Region Of The Paraná State, Brazil. *Acta Scientiarum* **2014**, *36*, 239–245.

Mandal, B.; Majumder, B.; Adhya, T. K.; Bandopadhyay, P. K.; Gangopadhyay, A.; Sarkar, D.; Kundu, M. C.; Gupta, S.,; Choudhury, G. C.; Hazra, G. C.; Kundu, S.; Samantaray, R. N.; Misra, A. K. Potential of Double-Cropped Rice Ecology to Conserve Organic Carbon Under Subtropical Climate. *Glob. Change Biol.* **2008**, *14*, 1–13.

McClaugherty, C.; Aber, J.; Melillo, J. The Role of Fine Roots in the Organic Matter and Nitrogen Budgets of Two Forested Ecosystems. *Ecology* **1982**, *63*, 1481–1490.

Nath, A. J.; Brahma, B.; Silesha, G. W.; Das, A. K.; Impact of Land Use Changes on the Storage of Soil Organic Carbon in Active and Recalcitrant Pools in a Humid Tropical Region of India, Science of The Total Environment, 2017, 624, 908–917.

Nelson, D. W.; Sommers, L. E. Total Carbon, Organic Carbon and Organic Matter. In *Methods of Soil Analysis, Part 2 Chemical and Microbiological Properties;* Page, A. L., Miller, R. H., Keeney, D. R., Eds., 2nd Ed.; The American Society of Agronomy and Soil Science Society of America: Madison, Wisconsin, USA, 1982; pp 539–579.

North Eastern Development Finance Corporation Ltd. North Eastern Data Bank 2016; NEDFi House, G. S. Road: Assam, India, 2016. http://databank.nedfi.com/content/assam (accessed Oct 15, 2017).

Pathak, K.; Nath, A. J.; Sileshi, G. W.; Lal, R.; Das, A. K. Annual Burning Enhances Biomass Production and Nutrient Cycling in Degraded *Imperata* Grasslands. *Land Degrad. Dev.* **2017**.

Paul, E. A.; Collins, H. P.; Leavitt, S. W. Dynamics of Resistant Soil Carbon of Midwestern Agricultural Soils Measured by Naturally Occurring 14C Abundance. *Geoderma* **2001**, *104*, 239–256.

Publicover, D. A.; Vogt, K. A. A Comparison of Methods for Estimating Forest Fine Root Production with Respect to Sources of Error. *Can. J. For. Res.* **1993**, *23*, 1179–1186.

Robertson, W. K.; Pope, P. E.; Tomlinson, R. T. Sampling Tool for Taking Undisturbed Soil Cores. *Soil Sci. Soc. Am. Proceed.* **1974**, *38*, 855–857.

Saha, D.; Kukul, S. S.; Sharma, S. Landuse Impacts on SOC Fractions and Aggregate Stability in Typicustochrepts of Northwest India. *Plant Soil* **2011**, *339*, 457–470.

Sarkar, D. C.; Meitei, B.; Baisya, L. K.; Das, A.; Ghosh, S.; Khumlo, L. C.; Rajkhowa, D. Potential of Allow Chronosequence in Shifting Cultivation to Conserve Soil Organic Carbon in Northeast India. *Catena* **2015**, *135*, 321–327.

Singh, S.; Dadhwal, V. K. *Manual on Spatial Assessment of Vegetation Carbon Pool of India*; Indian Institute of Remote Sensing (National Remote Sensing Centre), ISRO, Department of Space, Government of India: Dehradun, 2009.

Soil Survey Staff: Keys to Soil Taxonomy, 12th Ed.; USDA-Natural Resources Conservation Service: Washington DC, 2014.

Spohn, M.; Klaus, K.; Wanek, W.; Richter, A. Microbial Carbon Use Efficiency and Biomass Turnover Times Depending on Soil Depth-Implications for Carbon Cycling. *Soil Biol. Biochem.* **2016,** *96,* 74–81.

Straaten, O. V.; Corre, M. D.; Wolf, K.; Tchienkoua, M.; Cuellar, E.; Matthhews, R. B.; Veldkamp, E. Conversion of Lowland Tropical Forests to Tree Cash Crop Plantations Loses up to One-Half of Stored Soil Organic Carbon. *PNAS* **2015,** *112,* 9956–9960.

Tan, Z. H.; Zhang, Y. P.; Song, Q. H.; Lui, W. J.; Deng, X. B.; Tang, J. W.; Deng, Y.; Zhou, W. J.; Yang, L. Y.; Yu, G. R.; Sun, X. M.; Liang, N. S. Rubber Plantations Act as Water Pumps in Tropical China. *Geophys. Res. Lett.* **2011,** *38,* L24406.

The Rubber Board. *Rubber Growers Guide 2011*; Government of India: Kottayam, Kerala, India, 2011.

Vadivelu, S.; Sen, T. K.; Bhaskar, B. P.; Baruah, U.; Sarkar, D.; Maji, A. K.; Gajbhiye, K. S. *Soil Series of Assam,* NBSS Publ. No.101; NBSS & LUP: Nagpur, 2004; p 229.

Vesterdal, L.; Schmidt, I. K.; Callesen, I.; Nilsson, L. O.; Gundersen, P. Carbon and Nitrogen in Forest Floor and Mineral Soil Under Six Common European Tree Species. *For. Ecol. Manage.* **2008,** *255,* 35–48.

Walkley, A.; Black, I. A. An Examination of Degtjareff Method for Determining Soil Organic Matter and a Proposed Modification of the Chronic Acid Titration Method. *Soil Sci.* **1934,** *37,* 29–38.

Wang, Q.; Wang, S. Soil Organic Matter Under Different Forest Types in Southern China. *Geoderma* **2007,** *142,* 349–356.

Wang, Q.; Xiao, F.; He, T.; Wang, S. Responses of Labile Soil Organic Carbon and Enzyme Activity in Mineral Soils to Forest Conversion in the Subtropics. *Ann. For. Sci.* **2013,** *70,* 579–587.

Wauters, J. B.; Coudert, S.; Grallien, E.; Jonard, M.; Ponette, Q. Carbon Stock in Rubber Tree Plantations in Western Ghana And Matogrosso (Brazil). *For. Ecol. Manage.* **2008,** *255,* 2347–2361.

Wieder, R. K.; Lang, G. E. A Critique of the Analytical Methods Used in Examining Decomposition Data Obtained From Litter Bags. *Ecology* **1982,** *63,* 1636–1642.

Wigati, S.; Maksudi, M.; Latief, A. Analysis of Rubber Leaf (*Hevea brasiliensis*) Potency as Herbal Nutrition for Goats. In *Proceedings of the 16th AAAP Animal Science Congress,* Vol. II, Yogyakarta, 10–14 November, 2014; Yogyakarta (Indonesia): Gadjah Mada University, 2014; pp 497–500.

Wei, X. R. Shao, M. G.; Gale, W.; Li, L. H. Global Pattern of Soil Carbon Losses due to the Conversion of Forests to Agricultural Land. *Sci. Rep.* **2014,** *4,* 4062.

Zhang, M.; Fu, X. H.; Feng, W. T.; Zou, X. Soil Organic Carbon in Pure Rubber and Tea-Rubber Plantations in South-Western China. *Trop. Ecol.* **2007,** *48* (2), 201–207.

Ziegler, A. D.; Fox, J. M.; Xu, J. The Rubber Juggernaut. *Science* **2009,** *324,* 1024–1025.

CHAPTER 5

Ecosystem Carbon Sequestration

5.1 INTRODUCTION

Combating greenhouse gas (GHG) emission through reducing sources or enhancing sinks has been the priority theme of global research since mid-1990s. Among the GHGs, increase in carbon-di-oxide (CO_2) in the atmosphere through anthropogenic activities has been the prime cause of the global warming (Vashum and Jayakumar, 2012). Since direct CO_2 emission from land use change (LUC) alone contributes ~10% of total anthropogenic emission (Le Quere et al., 2016), it is one of the most important human-driven anthropogenic sources of atmospheric CO_2 (IPCC, 2014). Therefore, understanding of the changes in pools and fluxes of carbon (C) in soil and vegetation in the terrestrial ecosystems have attracted the attention of scientific community (Sarkar et al., 2015) to advance the understanding of climate change adaptation/mitigation. Tropical forests are the major source/sink of C (Wei et al., 2014) due to their strong (46%) impact on the global terrestrial C cycle (Bloom et al., 2016). In addition, a small change in tropical forests may strongly perturb the global C cycle while also jeopardizing several ecosystem services or ESs (Polasky et al., 2011). The annual C emission from deforestation related LUC is 0.9 Pg Cyear^{-1} (1 Pg = 1 gigaton = 10^{15} g = 1 million metric ton) compared with 9.0 Pg Cyear^{-1} for the decade 2005–2014 (Le Quere et al., 2016). Fossil fuel combustion and LUC, mostly in the tropical zones (Houghton and Goodale, 2004), have increased atmospheric CO_2 concentration to > 400 ppm in 2015 (Betts et al., 2016), and it is projected to exceed 500 ppm by 2050 (Ciais et al., 2013; WMO, 2016). Furthermore, rapid increase of atmospheric CO_2 concentration will increase earth surface temperature and sea level rise (IPCC, 2007 [WG-1]). Therefore, to combat these hazardous effects of climate change the United Nations Framework Convention on Climate Change (UNFCCC) formulated the Reduction Emission from Deforestation and Forest Degradation (REDD) in 2007, which was further enacted as REDD+ in 2010 (UNFCCC, 2008) to conserve and manage the total 2015 Pg of global terrestrial C stock (Jankte et al., 2016). Since the C in tropical forests are stored in different pools (vegetation and soil) and their accumulation rate and sink capacity vary with climatic variations, vegetative cover and intensity of human interventions (Lal, 2008; Lawrence et al., 2007), the LUC from native land can be a C sink or source (Zhang et al., 2015). Hence, assessing C stored in different pools under tropical land uses can advance scientific understanding of LUC and the terrestrial C cycle. In tropical forests, 67% of the ecosystem C (C in vegetation and in soil) resides in vegetation and 33% in soil (Ngo et al., 2013). However, the potentiality of such ecosystems to store C in vegetation and soil is influenced by the vegetative cover, climate, species, management practices, and human interventions (Ngo et al., 2013; Dung et al., 2016). The LUC from primary to secondary forests decreases both vegetation and soil C stocks (Ngo et al., 2013; Ogle et al., 2005) and degrades soil properties (Saha et al., 2011; Abera and Wolde-Meskel, 2013). Loss of biodiversity due to LUC from forest to plantations or grasslands degrades native vegetative structure and lowers the groundwater table (Ahrends et al., 2015), which further affect the ecosystem's C dynamics (De Camargo et al., 1999). Effect of LUC on either vegetation or soil C stock has been studied widely for diverse ecosystems (Fan et al., 2016; Murty et al., 2002; Wei et al., 2014; Hombegowda et al., 2016). However, the ecosystem C stock and its change with LUC have not been adequately studied for predominant land uses practiced in northeast India (NEI), which is an important ecoregion of the country.

In the present study, ecosystem C stock and sequestration rate were studied with respect to progressive and retrogressive changes in land uses. Progressive LUC was considered when any degraded land/forest was restored and conserved through developing commercial plantation or agroforestry systems. Similarly, when a natural forest was converted into any commercial plantation or severely degraded through anthropogenic perturbation, it was regarded as retrogressive changes in land uses. Therefore, the objective of this study was to assess the impacts of LUC and management on ecosystem C stock and sequestration potential for predominant land uses in NEI. The overall goal is to advance the scientific knowledge and use it to minimize the C emissions from LUC. The hypotheses tested were: (1) ecosystem C is lost through retrogressive changes in land uses and (2) restoration of degraded forests through traditional agroforestry systems can restore the ecosystem C to a level equivalent to that under a natural forest.

5.2 METHODS OF DATA COLLECTION

The present study was conducted in the Barak Valley (BV) region of Assam, NEI. The study area falls within the range of the Himalayan foothills and the Barak River Basin. The BV, in conjunction with Cachar, Hailakandi, and Karimganj administrative districts, covers an area of ~7000 km^2 (NEDFCL, 2016), and is characterized by the tropical humid climate. The BV region has the mean annual precipitation of 3500 mm, temperature range of 13–37°C and relative humidity of 93.5% (NEDFCL, 2016). The dominant soil of the region is classified as the Barak series, which is fine, mixed, hyperthermic family of *Aeric Endoaquepts* (USDA, 1998).

The ecosystem C stock, comprising biomass and soil, was measured for six predominant land uses of NEI (Roy & Bezbaruah, 2002). These LUCs were: (1) natural forest (NF), (2) disturbed forest (DF), (3) rubber (*Hevea brasiliensis*) plantation (RP), (4) Areca (*Areca catechu*) plantation (ArP), (5) pan (*Piper betle*) jhum (slash and mulching) agroforestry (PB), and (6) *Imperata* grassland (IG). The criteria for selection of NF and DF were the forest patch size, sapling density, tree density, and basal area of the forest (Majumdar and Datta, 2016). The strategy of selecting the representative six land uses was to assess changes in ecosystem C stock with progressive changes in land uses, namely (1) IG to RP/ArP or (2) DF to PB and with retrogressive changes in land uses namely (3) NF to DF to IG. Details of the predominant land use and their progressive and retrogressive changes are presented in Table 5.1 and Figure 5.1.

TABLE 5.1 The Details of the Predominant Land Uses Selected for the Study.

Land uses	Coordinates	District	Age (years)	Dominant species	Management details
ArP	24°48'02" N, 092°32'02" E	Karimganj	30	Monoculture of *Areca catechu* palm trees	Plantation space ~2 m between two palm trees. No fertilizers application

TABLE 5.1 *(Continued)*

Land uses	Coordinates	District	Age (years)	Dominant species	Management details
RP	24°36'02" N, 092°23'08" E	Karimganj	30	Monoculture of *Hevea brasiliensis* trees	Planting space varied from 3 to 4 m in between two trees. Fertilizer is applied at the initial stage of plantation development
PB	24°48'14" N, 092°33'24" E	Hailakandi	30	*Oroxylum indicum, Artocarpus chama, Mangifera indica, Areca catechu, Syzygium cumini, Musa sp.*, and *Terminalia chebula*	Production of beetle leaves is the prime reason for management of such agroforestry. Slash and mulching of woody tree leaves is practised annually
IG	24°41'44" N, 092°40'49" E	Cachar	30	*Imperata cylindrica*	Harvest of aboveground biomass and thereafter, burning the residue is practised annually
NF	24°40'57" N, 092°46'41" E	Cachar	Not known	*Cynometra polyandra, Ficus hispida, Litsea sp., Lindera sp., Artocarpus chama, Syzygium cumini,* and *Oroxylum indicum,* etc.	Naturally growing forest and represents tropical wet evergreen forest (Champion and Seth, 2005)
DF	24°49'03" N, 092°42'18" E	Cachar	30	*Memecylon celastrinum, Ficus hispida, Dysoxylum gobara, Vitex altissima, Syzygium cumini, Palaquium polyanthum, Artocarpus chama,* and *Pterospermum lancifolium*	Naturally growing forest and represents tropical wet evergreen forest. However, indiscriminate felling of trees for fuel wood has severely degraded the forest

Ecosystem Carbon Sequestration

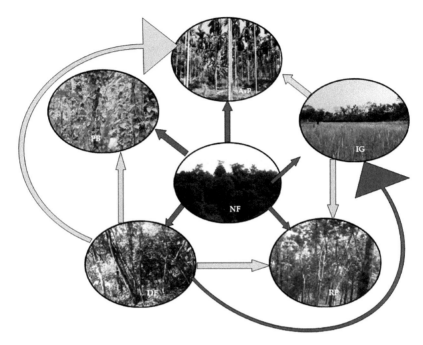

FIGURE 5.1 (See color insert.) Schematic presentation of progressive and retrogressive land use changes. Green color arrow denotes progressive changes and red color retrogressive changes.

5.2.1 Plot Selection for Biomass Sampling

For each of the five land uses (ArP, RP, PB, NF, and DF), four 250 m × 250 m quadrants were selected following the ISRO-GBP/NCP-VCP protocol (Singh and Dadhwal, 2009). The quadrants were chosen on representative sites covering dense to sparse vegetation cover. For each land use and each 250 m × 250 m quadrant, four 0.1 ha plots (31.62 m × 31.62 m) were laid out at four corners of the quadrant. In each plot, all trees >10 cm girth at the breast height (at 1.37 m from the base) were measured. For RP, the girth at the breast height was measured at 200 cm from the ground to avoid the tapping artifact (Brahma et al., 2016). Thus, all observations were made on a total of 16 plots, each of 0.1 ha area. Tree density (ha^{-1}) was calculated from the expanded values of each 0.1 ha plots by multiplying with 10 and averaged. The basal area (m^2 ha^{-1}) of each plot was calculated from the expanded values of each stem into hectare by multiplying the stem density and averaged. Height of each tree was measured using Nikon Laser Forestry Pro (serial no. 080753). The aboveground biomass (AGB) stock for NF, DF, and PB was estimated by using the biomass model of Brown et al. (1997), and the default values of the belowground biomass (BGB) were used to estimate the root biomass of tropical moist deciduous forests (Mokany et al., 2006), both were expressed on per hectare basis. However, direct harvesting of trees and extraction of roots were done to estimate the AGB and BGB stocks for RP and ArP. Based on the circumference range (60–120 cm) of the 30 year old RP, six major circumference classes (60–70 cm, 70–80 cm,…110–120 cm) were identified, and 10 trees from each circumference class were harvested. Thus, a total of 60 trees were harvested to assess the biomass for RP. For ArP trees, height is a better measure of biomass than the diameter (Frangi and Lugo, 1985; FAO, 2010). Therefore, based on height range (4–18 m), seven dominant height classes (4–6 m, 6–8 m…16–18 m) were identified, and five trees were harvested for each height class. Therefore, a total of 35 trees were harvested to generate biomass data for ArP. For estimating the BGB in RP and ArP, coarse roots (>2.5 mm) were excavated after felling the tree (Brahma et al., 2016). Roots were collected from the area of 1 m radius around the stem to 1 m depth. Fresh weights of all the plant parts (both AGB and BGB) were measured in the field. Sub-samples (500 g) were brought to the laboratory and oven dried at 70°C to the constant weight. The mean oven dry weight was multiplied with tree density in each plot and scaled up to assess the biomass per hectare for RP and ArP. For IG, the vegetation was harvested manually to obtain the biomass (AGB) data (Anderson and Ingram, 1993; Oliveras et al., 2014). Twenty evenly distributed 1 m^2 quadrant were established in the study site during December 2015, when plants were senesced, and the biomass was harvested at the ground level. The aboveground parts were weighed in field and sub-samples were subsequently oven dried at 70°C to the constant weight. *Imperata* is a shallow-rooted plant (Pathak et al., 2017) and therefore below ground parts were extracted to 30 cm depth following the core method (Anderson and Ingram, 1993; Yazaki et al., 2004). These samples were brought to the laboratory, washed thoroughly to remove soil, weighed, and oven dried at 70°C to the constant weight. The oven dry-weight for ArP, RP, and IG was scaled up to assess the biomass per hectare. Carbon density for each of the land uses was estimated by using the default fraction of biomass C at 47%

of the total weight, following the procedure by Andreae and Merlet (2001) and McGroddy et al. (2004).

5.2.2 *Soil Sample Collection and Preparation*

Soil samples were collected from each of the land uses during February and March, 2016. Two soil profiles (1 m × 1 m × 1 m) were dug at randomly chosen sites within each of the 250 m × 250 m (for ArP, RP, PB, NF, and DF) and 1 m × 1 m (for IG) sized quadrant used for vegetation sampling. Thus, a total of eight profiles were dug for collecting soil samples for each land use. Soil samples were collected using cores of 5.6 cm diameter and 30 cm length from 0–10, 10–30, and 30–100 cm depths. Core samples were collected in triplicate by inserting the core horizontally up to 10 cm for each depth. In addition to core, bulk soil samples (a total of 72, 8 profiles × 3 depths × 3 samples from each depth) were also obtained for each land use, air-dried and sieved through 100 μ mesh for assessment of soil organic carbon (SOC) concentration by the wet oxidation method (Walkley and Black, 1934; Nelson and Sommers, 1982). The SOC stocks (Mg ha^{-1}) for each depth under different land uses were computed following the method of Blanco-Canqui and Lal (2008).

The SOC stocks for different depths (0–10, 10–30, and 30–100 cm) were summed up to represent SOC stock for 0–100 cm depth. The ratio of SOC: ecosystem C was computed by dividing SOC stock (up to 100 cm) by ecosystem C stock for all the land uses to delineate the proportion of SOC to ecosystem C stock. Soil texture of the composite soil sample was determined by the hydrometer method (Gee and Bauder, 1986). Soil BD was estimated using the method described by Robertson et al. (1974).

The ecosystem C sequestration rate was computed for the following progressive LUC and management: (1) DF into ArP, (2) DF into RP, (3) DF into PB, (4) IG to ArP, and (5) IG to RP because age of such systems (30 years) was known to us. For calculating ecosystem C sequestration rate (Mg ha^{-1} year^{-1}) after the progressive LUC, the change in ecosystem C stock was divided by the respective age (30 years) of the land use. Because the exact age of forest is not known, it was not possible to compute the ecosystem C sequestration rate for the retrogressive LUC and management for: (1) NF to ArP, (2) NF to RP, (3) NF to PB, (4) NF to DF, and (5) NF to IG.

Shapiro–Wilk tests (Shapiro and Wilk, 1965) and Levene's tests (Gastwirth et al., 2009) were performed to test the normality distribution and homogeneity of variance of the SOC content and BD of different depths under different land uses. One way ANOVA was tested at 95% confidence level to know the significant differences in SOC content and BD values of different depths among all the land uses (Zar, 1999). Tukey HSD (honestly significant difference) post hoc test was performed (Pereire et al., 2014), if the one-way ANOVA resulted in any significant differences ($P < 0.05$). All statistical calculations were done using MS Excel 2003 and IBM SPSS Statistics 21 software.

5.3 LAND USE CHARACTERISTICS FOR DIVERSE ECOSYSTEM TYPES

The highest stem density among all the land uses was measured for ArP (1560 ha^{-1}) and the lowest under DF (404 ha^{-1}). The one-way ANOVA shows that stem density differed significantly among the land uses in NEI ($P < 0.05$) and the Tukey HSD test showed significant differences between ArP and PB (Table 5.2). The basal area was the highest under RP (29.82 m^2 ha^{-1}) and the lowest under ArP (16.94 m^2 ha^{-1}; Table 5.2). However, the Tukey HSD test signifies no significant differences in basal area among ArP, NF, and DF ($P > 0.05$). The soil textural class was sandy loam to sandy clay loam for all the land uses (Table 5.2). The highest AGB was observed under NF (293.6 Mg ha^{-1}) followed by that under RP (269.1 Mg ha^{-1}). In contrast, the highest BGB was observed under PB (65.1 Mg ha^{-1}) followed by that under NF (62.5 Mg ha^{-1}; Fig. 5.2). Soil BD for 0–10 cm depth was the highest under ArP (1.49 Mg m^{-3}) and the lowest under PB (1.02 Mg m^{-3}; Table 5.3). The highest and lowest BD for 10–30 cm soil depth was under IG (1.59 Mg m^{-3}) and DF (1.16 Mg m^{-3}) respectively, whereas for the 30–100 cm soil depth the highest and the lowest BD was found under DF (1.55 Mg m^{-3}) and ArP (1.32 Mg m^{-3}; Table 5.3). The highest SOC content for 0–10 cm soil depth was observed under PB (18.50 g kg^{-1}) followed by that under NF (17.90 g kg^{-1}) and the lowest was under IG (12.10 g kg^{-1}; Table 5.3). The highest SOC content for the successive lower depths (10–30 cm and 30–100 cm) for different land uses were observed for PB (11.20 g kg^{-1}) and NF (8.00 g kg^{-1}), respectively (Table 5.3). The lowest SOC content for 10–30 and 30–100 cm soil depths were observed under IG (6.90 and 5.70 g kg^{-1}, respectively; Table 5.3). The BD and SOC data were considered as normally distributed after Shapiro–Wilk test (at $P > 0.05$ confidence level). Furthermore, homogeneity of variances of the data was reflected on the Levene's test ($P > 0.05$). One-way ANOVA ($P < 0.05$) signifies that the BD and SOC content for the different depths are significantly different among the land uses (Table 5.3). Soil BD and SOC content in 0–10 and 10–30 cm depth were negatively correlated when data for all land uses were lumped.

TABLE 5.2 Land Use Characteristics for Six Predominant Land Uses of Northeast India.

Land uses	Stem density (ha^{-1})	Basal area (m^2 ha^{-1})	Soil texture class
ArP	1560a (33)	16.94a (0.35)	Sandy loam
RP	714b (36)	29.82b (1.54)	Sandy clay loam
PB	1350a (53)	27.89b (1.09)	Sandy clay loam
IG	NA	NA	Sandy loam
NF	961bc (70)	23.79a (1.73)	Sandy clay loam
DF	404bd (47)	21.07a (2.43)	Sandy loam
F value	89.47	10.71	
P value	0.000	0.000	

Note: Values are mean and standard errors are given in the parenthesis. Results of one way ANOVA at 95 % confidence level has been given in the lower portion of the table. Same letter signifies that there is no significant difference between the values of different land uses. ArP, *Areca* plantation; DF, disturbed forest; IG, *Imperata* grassland; NA, not applicable; NF, natural forest; PB, *Piper betle* agroforestry; RP, rubber plantation.

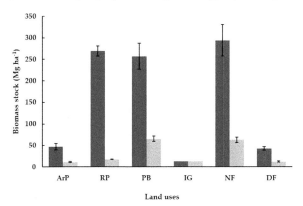

FIGURE 5.2 (See color insert.) Biomass stock (above and below ground) for predominant land uses of northeast India. Values are mean ± standard error. Blue bars: aboveground biomass stock and red bars: belowground biomass stock. ArP, *Areca* plantation, DF, disturbed forest; IG, *Imperata* grassland; NF, natural forest; PB, *Piper betle* agroforestry; RP, rubber plantation.

TABLE 5.3 Bulk Density (BD) and Soil Organic Carbon (SOC) Content of Predominant Land Uses of Northeast India.

Land uses	BD (Mg m⁻³)			SOC (g kg⁻¹)		
	0–10 cm	10–30 cm	30–100 cm	0–10 cm	10–30 cm	30–100 cm
ArP	1.49a (0.01)	1.51a (0.08)	1.32a (0.06)	12.80a (1.3)	7.65a (0.8)	5.85a (0.4)
RP	1.28a (0.07)	1.46a (0.02)	1.33a (0.04)	17.55b (0.7)	8.80ac (0.3)	5.80a (0.3)
PB	1.02b (0.01)	1.46a (0.07)	1.39a (0.02)	18.50 b (2)	11.20bc (0.4)	6.60ab (0.4)
IG	1.25a (0.05)	1.59a (0.02)	1.53b (0.07)	12.10a (1)	6.90a (0.6)	5.70a (0.3)
NF	1.20b (0.08)	1.34b (0.05)	1.53b (0.03)	17.90 b (4)	9.70ac (2)	8.00b (0.5)
DF	1.21a (0.04)	1.16b (0.04)	1.55b (0.01)	10.65a (0.9)	7.85a (0.4)	5.73a (0.2)
F value	8.85	8.46	6.61	23.37	11.91	6.45
P value	0.010	0.011	0.020	0.001	0.005	0.021

Note: Values are mean and standard errors are given in the parenthesis. Results of one way ANOVA at 95% confidence level has been given in the lower portion of the table. Same letter signifies that there is no significance difference between the observations of different land uses. ArP, *Areca* plantation, BD, bulk density; DF, disturbed forest; IG, *Imperata* grassland; NF, natural forest; PB, *Piper betle* agroforestry; RP, rubber plantation; SOC, soil organic carbon content.

5.4 BIOMASS, SOIL, AND ECOSYSTEM C STOCK

Vegetation C stock (aboveground + belowground) was the highest under NF (167.37 Mg ha⁻¹) and follows the trend: NF > PB > RP > ArP > DF > IG (Table 5.4). The estimated SOC stock up to 100 cm depth ranges from 96.2 (ArP) to 133.1 Mg ha⁻¹ (NF; Table 5.4). The estimated ecosystem C stock ranges from 110.4 (IG) to 300.5 (NF) Mg ha⁻¹ and follows the trend: NF > PB > RP > ArP > DF > IG (Table 5.4). The ratio of SOC stock to ecosystem C stock was the highest for IG (89%) followed by that for DF (79%), and the lowest was for PB and RP (43%; Table 5.4).

TABLE 5.4 Biomass C, SOC, and Ecosystem C Stock of Predominant Land Uses of Northeast India.

Land uses	AGBCS	BGCS	VCS	SOCS (0–100 cm)	ECS	SOC: EC
	(Mg ha^{-1})	(Mg ha^{-1})	(Mg ha^{-1})	(Mg ha^{-1})	(Mg ha^{-1})	(%)
ArP	22.04a (0.46)	5.08a (0.11)	27.12a (0.57)	96.18a (8.07)	123.30a (6.63)	78
RP	126.48b (6.53)	8.01a (0.41)	134.49b (6.94)	101.95b (4.98)	236.43b (10.39)	43
PB	120.74b (4.70)	30.57b (1.19)	151.31b (5.89)	115.85c (4.37)	267.17b (8.50)	43
IG	6.15a (0.39)	6.43a (0.33)	12.58a (0.66)	97.81a (5.45)	110.39a (5.84)	89
NF	138.01b (10.03)	29.36b (2.13)	167.37a (12.16)	133.08d (7.01)	300.45bc (14.87)	44
DF	19.95a (2.30)	5.46a (0.63)	25.41a (2.93)	93.55a (6.16)	118.95a (2.19)	79
F value	134.76	136.28	128.48	12.00	89.18	
P value	0.000	0.000	0.000	0.000	0.000	

Note: Values are mean and standard errors are given in the parenthesis. Results of one way ANOVA at 95% confidence level has been given in the lower portion of the table. Same letter signifies that there is no significance difference between the observations of different land uses. AGBCS, aboveground biomass carbon stock; ArP, *Areca* plantation, BGBCS, belowground biomass carbon stock; DF, disturbed forest; ECS, ecosystem carbon stock; IG, *Imperata* grassland, NF, natural forest; PB, *Piper betle* agroforestry; RP, rubber plantation; SOCS, soil organic carbon stock; VCS, vegetation carbon stock.

5.5 CHANGES IN BIOMASS, SOIL, AND ECOSYSTEM C STOCK

With progressive LUC, the vegetation C stock under ArP and PB was 7% and 495% more than that under DF, respectively (Table 5.5). Similarly, the vegetation C stock was 116% and 969% more when IG was converted into ArP and RP, respectively (Table 5.5). In contrast, with retrogressive LUC, vegetation C under PB and IG were 10% and 92% less than that in the NF, respectively (Table 5.5).

With the progressive LUC from DF to PB/RP, an increase in SOC stock in surface layer (0–10 cm) by 47% and 75 % was observed over that under DF. For 10–30 cm soil depth, the increase in SOC stock was as high as 79% for PB, in comparison with that under DF. However, in lower soil depth (30–100 cm), 12% decline in SOC stock in ArP was observed over that under DF (Table 5.5). For 0–100 cm depth, the soil under progressive LUC shows 3% and 24% increase in SOC stock under ArP and PB, respectively, over that under DF. However, with progressive changes in land uses from IG, there was only a marginal increase (4%) in SOC stock (up to 100 cm) under RP (Table 5.5). With retrogressive LUC, the SOC stock for the 0–10 cm soil depth shows marginal gain under RP (1.1 Mg ha^{-1}) and substantial loss under DF (8.6 Mg ha^{-1}) compared with that under NF. The gain under RP and loss under DF were computed at 5% and 40%, respectively, in comparison with that under NF. There were 26% increase in SOC stock under PB and 30% loss under DF over that under NF in 10–30 cm soil depth. For lower soil depth (30–100 cm), the loss of SOC stock was as high as 37% for RP over that under NF. Considering the cumulative stock for 0–100 cm, changes in SOC stock indicate 13% and 30% loss for PB and DF, respectively, over that under NF (Table 5.5).

The retrogressive LUC from NF to PB and IG shows 11% and 63% less ecosystem C stock than that under NF, respectively (Table 5.5; Fig. 5.3). However, a gain in ecosystem C stock was observed for all the progressive land uses considering the DF as the control land use. The sequence in ecosystem C gain was in the order: PB (125%) > RP (99%) > ArP (4%). Furthermore, the gain in ecosystem C stock was 114% and 12% for RP and ArP, respectively, with IG as control land use (Table 5.5; Fig. 5.3).

TABLE 5.5 Changes in Biomass Carbon, SOC, and EC after Progressive and Retrogressive Changes in Predominant Land Uses in Northeast India.

Land use changes and management	Biomass C changes			SOCS changes			ECS changes
	AGBCS	BGBCS	VCS	0–10 cm	10–30 cm	30–100 cm	0–100 cm
	(Mg ha^{-1})	(Mg ha^{-1})	(Mg ha^{-1})	(Mg ha^{-1})	(Mg ha^{-1})	(Mg ha^{-1})	(Mg ha^{-1})
Progressive land use changes and management							
DF to ArP	2.09 (+10)	−0.38 (−7)	1.71 (+7)	6.22 (+48)	4.84 (+27)	−7.25 (−12)	2.63 (+3)	4.35 (+4)
DF to RP	106.53 (+534)	2.55 (+47)	109.08 (+429)	9.69 (+75)	7.44 (+41)	−8.43 (−14)	8.40 (+9)	117.48 (+99)

TABLE 5.5 *(Continued)*

Land use changes and management	Biomass C changes			SOCS changes			ECS changes
	AGBCS	BGBCS	VCS	0–10 cm	10–30 cm	30–100 cm	0–100 cm
	(Mg ha^{-1})	(Mg ha^{-1})	(Mg ha^{-1})	(Mg ha^{-1})	(Mg ha^{-1})	(Mg ha^{-1})	(Mg ha^{-1})
DF to PB	100.79 (+505)	25.12 (+460)	125.91 (+495)	6.02 (+47)	14.46 (+79)	2.12 (+3)	22.3 (+24)
IG to ArP	15.89 (+258)	−1.35 (−21)	14.54 (+116)	3.94 (+26)	1.11 (+5)	−5.80 (−10)	−1.63 (+2)
IG to RP	120.33 (+1957)	1.58 (+25)	121.91 (+969)	7.41 (+49)	3.72 (+17)	−6.99 (−12)	4.14 (+4)
Retrogressive land use changes and management							
NF to ArP	−115.97 (−84)	−24.28 (−83)	−140.25 (−84)	−2.38 (−11)	−2.90 (−11)	−30.67 (−36)	−36.94 (−28)	−177.17 (−59)
NF to RP	−11.53 (−8)	−21.35 (−73)	−32.88 (−20)	1.09 (+5)	−0.30 (−1)	−31.85 (−37)	−31.13 (−23)	−64.02 (−21)
NF to PB	−17.27 (−13)	1.22 (+4)	−16.05 (−10)	−2.58 (−12)	6.72 (+26)	−21.30 (−25)	−17.23 (−13)	−33.28 (−11)
NF to IG	−131.86 (−96)	−22.93 (−79)	−154.78 (−92)	−6.32 (−−29)	−4.01 (−15)	−24.86 (−29)	−35.27 (−27)	−190.06 (−63)
NF to DF	−118.06 (−86)	−23.90 (−81)	−141.96 (−85)	−8.60 (−40)	−7.74 (−30)	−23.42 (−27)	−39.53 (−30)	−181.50 (−60)

Note: Values are mean and percentage gain (+) or loss (−) in biomass C, SOC, and ecosystem C are given in the parenthesis. ArP, *Areca* plantation, AGBCS, aboveground biomass carbon stock; BGBCS, belowground biomass carbon stock; DF, disturbed forest; ECS, ecosystem carbon stock; IG, *Imperata* grassland; NF, natural forest; PB, *Piper betle* agroforestry; RP, rubber plantation, SOCS, soil organic carbon stock; VCS, vegetation carbon stock.

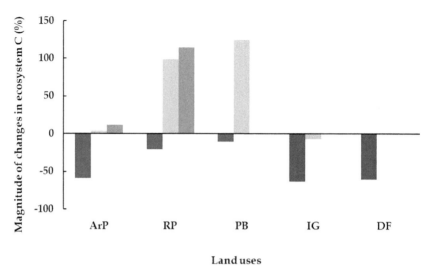

FIGURE 5.3 (See color insert.) Magnitude (%) of changes in ecosystem C over their respective control land use. Blue bars: magnitude of ecosystem carbon changes considering NF as control, red bars: considering DF as control, and green bars: considering IG as control land use. ArP, *Areca* plantation, DF, disturbed forest; IG, *Imperata* grassland; PB, *Piper betle* agroforestry; RP, rubber plantation.

5.6 VEGETATION, SOIL, AND ECOSYSTEM CARBON SEQUESTRATION

With progressive LUC, the highest vegetation C sequestration rate was computed at 4.2 Mg ha^{-1} year^{-1} when DF was converted into PB. Soil C sequestration rate was the highest (0.74 Mg ha^{-1} year^{-1}), when DF was converted into PB and the cumulative stock was considered to 100 cm depth. However, a marginal loss in SOC was recorded when IG was converted into ArP (−0.05 Mg ha^{-1} year^{-1}). The ecosystem C sequestration rate was the highest when DF was converted to PB (4.9 Mg ha^{-1} year^{-1}), and estimated at 4.2 Mg ha^{-1} year^{-1} when IG was converted into RP (Table 5.6).

TABLE 5.6 Biomass C, SOC, and Ecosystem C Sequestration Rate After Progressive Land Use Changes in Northeast India.

Land use changes and management	VC Seq	SOC Seq (0–100 cm)	EC Seq
	(Mg ha^{-1} yr^{-1})	(Mg ha^{-1} yr^{-1})	(Mg ha^{-1} yr^{-1})
DF to ArP	0.06	0.09	0.14
DF to RP	3.64	0.28	3.92
DF to PB	4.20	0.74	4.94
IG to ArP	0.48	−0.05	0.43
IG to RP	4.06	0.14	4.20

ArP: *Areca* plantation; DF, disturbed forest; EC Seq, ecosystem carbon sequestration; IG, *Imperata* grassland; PB, *Piper betle* agroforestry; RP, rubber plantation, SOC Seq, soil organic carbon sequestration; VC Seq, vegetation carbon sequestration.

5.7 LAND USE CHARACTERISTICS AND ITS IMPLICATIONS

The stem density in natural or plantation forest is mostly influenced by the intensity of anthropogenic interventions and management practices (Schall and Ammer, 2013). Since the DF of NEI involves the human-induced land use conversion of primary forests (Poffenberger et al., 2006), the lowest stem density (404 ha^{-1}) observed in the present study is attributed to deforestation. The highest stem density (1560 ha^{-1}) under ArP is due to reduced plantation spacing (~2 m) in comparison to other commercially grown tree species (Brahma et al., 2016). The stem density trend for different land uses was in the order: ArP > PB > NF > RP > DF. In accord with the variations in stem density, the basal area also differed among the land uses. The basal area of any ecosystem/land uses is influenced by species composition, tree size and growth pattern (Condes and Rio, 2015). The trend for basal area was in the order: RP > PB > NF > DF > ArP. The RP being a mature plantation (30 year old), the stem sizes distributed among circumference classes of 60–120 cm and resulted in the highest basal area (29.8 m^2 ha^{-1}) in comparison to those for other land uses. In contrast, the stem circumference class of ArP was mostly confined within the circumference range of 20–50 cm, leading to the lowest basal area of 16.94 m^2 ha^{-1}. Comparatively high basal area under PB compared with that under NF is in accord with the results of Nandy and Das (2013), also from a similar geographical region. Since, the land uses of the present study comprise a variety of plants (i.e., trees, palm, and grasses), comparing the petiole density and the total petiole area of *Imperata* grass to the stem density and basal area of trees are invalid comparisons. Therefore, the comparative statements for stem density and basal area are not presented separately for the IG.

5.8 CHANGES IN BIOMASS C STOCK THROUGH LUC AND MANAGEMENT

The total biomass C stock was the highest under NF (167.4 Mg C ha^{-1}) of which 83% (138 Mg ha^{-1}) was the AGB (Table 5.4). The default AGB C stock for an equivalent land use in South Asia is 138 Mg ha^{-1} (IPCC, 2014), and is comparable to that of the present study. The LUC from NF to other land uses show that PB agroforestry is more resilient system with only 9% less vegetation C than that under forest. The loss of vegetation C was as high as 83–85% when NF was converted into DF or ArP. Such a significant loss in vegetation C due to LUC into plantation/commercial crops has also been reported by Liao et al. (2010) and Fan et al. (2016). The highest loss of vegetation C (93%) observed for IG is comparable to 95–98% loss of vegetation C stock in grasslands of Indonesia in comparison with that under native forest (Palm et al., 2005).

5.9 CHANGES IN SOC STOCK THROUGH LUC AND MANAGEMENT

Under progressive LUC, the SOC stock followed an increasing trend for 0–10 and 10–30 cm depths. However, decrease in SOC stock was observed in 30–100 cm depth when DF and IG were converted to ArP/RP. This decline in SOC stock is attributed to decrease in BD in 30–100 cm depth for ArP and RP in comparison with that for 0–10 and 10–30 cm layers. The SOC stock up to 100 cm depth was the highest for NF (133.1 Mg ha^{-1}), and was in the order: NF > PB > RP > IG > ArP > DF. Soils at the equilibrium state (e.g., undisturbed or less disturbed forest soils) and those not prone to soil-related constraints (e.g., acid soils with high exchangeable Al^{3+} and H$^+$, low P and Ca^{2+}) contain the highest amount of SOC stock (Lal, 2004, 2005). Comparable ranges of SOC stocks (177 Mg ha^{-1}) have been reported for forests of Kerala, India (Saha et al., 2011) and other tropical regions (Chhabra et al., 2003). Traditionally managed agroforestry systems store more or equal SOC stock to that of tropical forests (Negash and Starr, 2015; Hombegowda et al., 2016) depending on the species diversity and management systems (Lenka et al., 2012). The SOC stock under traditional PB agroforestry system in the present study was estimated at 115.9 Mg ha^{-1}, which was only 13% lower than that under NF. The SOC stock of agroforestry systems in some regions of southern India was ~15% lower than that under the native forest (Hombegowda et al., 2016). Almost similar SOC stock among PB and NF land uses is attributed to the slash and mulch-based nutrient management system in the former. Under PB agroforestry system, management practices like pruning of host tree branches and decomposing these onsite, might have contributed to increase in soil organic matter (SOM). Conversion of forests into an arable land use reduces SOC stock by 25%–60% in the top 100 cm depth (Wei et al., 2014; Hombegowda et al., 2016). The SOC stock under RP in the present study was comparable to that of some soils in China (De Blecourt et al., 2013). The retrogressive change from NF to RP depletes SOC to 100 cm depth by 23%. In China, conversion of native forest into Chinese fir (*Cunninghamia lanceolata*) and rubber plantations decreased SOC stock by 19%–38 % compared with that in soil under native vegetation (Guillaume et al., 2015; Fan et al., 2016). The SOC stock to 100 cm depth in IG (97.8 Mg ha^{-1}) is in accord with that reported in other studies (Henderson et al., 2016), and is ~27% lower than that of NF. Studies from other parts of India also showed that grasslands contained the least (23% less) SOC stock in comparison with that under the native forests (Iqbal and Tiwari, 2016). LUC from NF to IG accelerated soil erosion because of a low or no vegetative cover following the annual harvest, and the biomass burning depleted the SOC stock. The lowest SOC stock to 100 cm depth was observed under DF (93.6 Mg ha^{-1}), which was 30% (39.8 Mg ha^{-1}) lower than that under NF. Similar results have been reported when native forests were converted into secondary forests in Colombia (Sierra et al., 2007).

5.10 CHANGES IN ECOSYSTEM C STOCK THROUGH LUC AND MANAGEMENT

The ecosystem C stocks under NF, RP, PB, and IG of NEI are comparable with those reported for rubber plantations, traditional agroforestry, and grassland ecosystems of South East Asia and other parts of the world (Sierra et al., 2007; Zeigler et al., 2012; Negash and Starr, 2015). With retrogressive change, the loss of ecosystem C was the lowest under PB and the highest under IG. These results are in accord with those reported by Jackson et al. (2002) and Henderson et al. (2016). The LUC from NF to DF caused 60% (181.5 Mg ha^{-1}) loss of the ecosystem C, which was 26% higher than that reported by Ngo et al. (2013). Conversion of NF to RP and ArP depleted the SOC stock by 21% (64.0 Mg ha^{-1}) and 59% (177.2 Mg ha^{-1}) of ecosystem C, respectively. A similar trend in loss of the ecosystem C by 37% as compared with that under native forest was estimated for Chinese fir plantations (Fan et al., 2016). The ecosystem C stock of PB being 11% less than that under NF suggests that traditional agroforestry management system accelerates biomass and soil C sequestration, which is similar to that for the NF.

The progressive LUC from DF to other land uses exhibited a net gain in ecosystem C stocks, except for that under IG. Conversion of DF to ArP/RP caused a net gain of ecosystem C by 4% (4.4 Mg ha^{-1}) and 99% (117.5 Mg ha^{-1}), respectively. Since, the LUC from DF to ArP does not significantly influence the biomass (+1.7 Mg ha^{-1}) and SOC stocks (+2.8 Mg ha^{-1}); therefore, the ecosystem C stocks under ArP and DF are similar. However, the significant gain in ecosystem C under RP is comparable with that of some pine plantations in China when established on soil under a degraded forest (Li et al., 2015). Furthermore, in the present study, the progressive conversion of IG into ArP/RP resulted into gain of ecosystem C by 12% and 114%, respectively (Table 5.5 and Fig. 5.3). Invasion by woody vegetation on grasslands resulted into a significant gain in ecosystem C by increasing both biomass and soil C stocks (Huang et al., 2007).

5.10.1 SOC: Ecosystem C Ratio

The ratio of SOC: ecosystem C was < 50% for NF, PB, and RP (44%, 43%, and 43%, respectively). These results are in accord with those for some tropical agroforestry and forestry systems (Negash and Starr, 2015). However, the ratio is distorted when forest is converted into plantation or grassland. For example, the ratio of SOC: ecosystem C stock was 78%, 79%, and 89% for ArP, DF, and IG, respectively. Therefore, retrogressive LUC stores proportionately more C in soil, and hence, makes such land use prone to depletion of the SOC stock through accelerated soil erosion resulting from low forest floor biomass and poor vegetative cover. Hence, stabilizing and managing traditional/improved agroforestry in NEI (Murthy et al., 2013) on degraded forest/grasslands can restore the SOC stock and set in motion a rapid succession and regeneration through revegetating the degraded landscape.

5.10.2 *Ecosystem C Sequestration and its Implication in Restoration of Degraded Lands*

The restoration of DF through ArP had an average rate of ecosystem C sequestration of 0.14 Mg ha^{-1} year^{-1}, but the rate was as high as 4.9 and 3.9 Mg ha^{-1} year^{-1} for restoration through PB and RP, respectively. Furthermore, the ecosystem C sequestration rate was 4.2 Mg ha^{-1} year^{-1} when IG was restored through RP. Therefore, restoration of degraded lands (namely, DF, IG) through RP and PB substantially enhances ecosystem C sequestration rate and simultaneously advances the process of climate change mitigation/adaptation. Land area of 1.5 M ha is under shifting cultivation in NEI (Roy et al., 2012), and these ecosystems are degraded physically, chemically, and biologically (Nath et al., 2016). Therefore, such degraded lands can be restored through conversion to RP and PB. Although ecosystem C sequestration rate of RP is consistent with that of PB, but the former has often been associated with diverse environmental issues including the loss of biodiversity (Warren-Thomas et al., 2015) and disruption of the hydrological cycle (Ziegler et al., 2009). Therefore, considering the perspective of environment management and conservation, PB agroforestry would be a better alternative to accelerate ecosystem C sequestration on degraded lands.

5.11 CONCLUSIONS

The data presented proves both hypotheses: (1) ecosystem C is lost through retrogressive changes in land uses and (2) restoration of degraded forests through traditional agroforestry system can achieve the ecosystem C stock equivalent to that under the NF. The data presented also support the following conclusions: (1) the magnitude of loss of ecosystem C through LUC from NF into any other land uses ranged from 11% to 63%, depending on the specific land use, (2) traditional PB agroforestry systems are highly resilient system, with only 11% loss of the ecosystem C than that under the NF, (3) the technique of slash and mulch as practiced under PB accelerates ecosystem C sequestration rate, and (4) LUC from NF to ArP/DF/IG distorts the SOC: ecosystem C ratio.

KEYWORDS

- **carbon sequestration**
- **progressive LUC**
- **retrogressive LUC**
- **plantation systems**
- **agroforestry systems**

REFERENCES

Abera, G.; Wolde-Meskel, E.; Soil Properties, and Soil Organic Carbon Stocks of Tropical Andosol Under Different Land Uses. *Open J. Soil Sci.* **2013**, *3*, 153–162.

Ahrends, A.; Hollingsworth, P. M.; Ziegler, A. D.; Fox, J. M.; Chen, H.; Su, Y.; Xu, J. Current Trends of Rubber Plantation Expansions May Threaten Biodiversity and Livelihoods. *Glob. Environ. Change* **2015**, *34*, 48–58.

Anderson, J. M.; Ingram , J. S. I. (Eds.). *Tropical Soil Biology and Fertility: A Handbook of Method;* CAB International, 1993.

Andreae, M. O.; Merlet, P. Emission of Trace Gases and Aerosols from Biomass Burning. *Glob. Biogeochem. Cycles* **2001**, *15*, 955–966.

Betts, R. A.; Jones, C. D.; Knight, J. R.; Keeling, R. F.; Kennedy, J. J. El Nino and a Record CO_2 Rise. *Nat. Climate Change* **2016**, *6*, 806–810.

Blanco-Canqui, H.; Lal, R. No-tillage and Soil-profile Carbon Sequestration: An On-farm Assessment *Soil Sci. Soc. Am. J.* **2008**, *72*, 693–701.

Bloom, A. A.; Exbrayat, J. F.; Velde, I. R. V. D.; Feng, L.; Williams, M. The Decadal State of the Terrestrial Carbon Cycle: Global Retrievals of Terrestrial Carbon Allocation, Pools, and Residence Times. *PNAS* **2016**, *113*, 1285–1290.

Brahma, B.; Nath, A. J.; Das, A. K. Managing Rubber Plantations for Advancing Climate Change Mitigation Strategy. *Curr. Sci.* **2016**, *110*, 2015–2019.

Brown, S. Estimating Biomass and Biomass Change of Tropical Forests: A Primer. In *FAO Forestry Paper;* Food and Agriculture Organization of the United Nations: Rome, 1997; p 66.

Chhabra, A.; Palria, S.; Dadhwal, V. K. Soil Organic Carbon Pool in Indian Forests. *For. Ecol. Manag.* **2003**, *173*, 187–199.

Ciais, P.; Sabine, C.; Bala, G.; Bopp, L.; Brovkin, V.; Canadell, J.; Chhabra, A.; DeFries, R.; Galloway, J.; Heimann, M.; Jones, C.; Quéré, C. L.; Myneni, R. B.; Piao, S.; Thornton,

P. Carbon and Other Biogeochemical Cycles. In *Climate Change 2013: The Physical Science Basis. Contribution of Working Group I to the Fifth Assessment Report of the Intergovernmental Panel on Climate Change*; Stocker, T. F.; Qin, D.; Plattner, G. K.; Tignor, M.; Allen, S. K.; Boschung, J.; Nauels, A.; Xia, Y.; Bex, V.; Midgley, P. M., Eds.; Cambridge University Press: Cambridge, United Kingdom and New York, NY, USA, 2013; pp 465–570. DOI:10.1017/CBO9781107415324.015, 2013.

Condes, S.; Del Rio, M. Climate Modifies Tree Interactions in Terms of Basal Area Growth and Mortality in Monospecific and Mixed *Fagus sylvatica* and *Pinus sylvestris* Forests. *Euro. J. For. Res.* **2015**, *134*, 1095–1108.

De Blecourt, M.; Brumme, R.; Xu, J.; Corre, M. D.; Veldkamp, E. Soil Carbon Stocks Decrease Following Conversion of Secondary Forests to Rubber (*Hevea brasiliensis*) Plantations. *PLoS One* **2013**, *8*, 69357.

De Camargo, P.; Trumbore, S.; Martinelli, L.; Davidson, E.; Nepstad, D.; Victoria, R. Soil Carbon Dynamics in Regrowing Forest of Eastern Amazonia. *Glob. Chang. Biol.* **1999**, *5*, 693–702.

Dung, L. V.; Tue, N. T.; Nhuan, M. T.; Omori, K. Carbon Storage in a Restored Mangrove Forest in Can Gio Mangrove Forest Park, Mekong Delta, Vietnam. *For. Ecol. Manag.* **2016**, *380*, 31–40.

Fan, S.; Guan, F.; Xu, X.; Forrester, D. I.; Ma, W.; Tang, X. Ecosystem Carbon Stock Loss After Land Use Change in Subtropical Forests in China. *Forests* **2016**, *7*, 142.

FAO. *Global Forest Resource Assessment*; Food and Agricultural Organization of The United Nations (FAO): Rome, 2010.

Frangi, J. L.; Lugo, A. E. Ecosystem Dynamics of a Subtropical Floodplain Forest. *Ecol. Monographs* **1985**, *55*, 351–369.

Gastwirth, J. L.; Gel, Y. R.; Miao, W. The Impact of Levene's Test of Equality of Variances on Statistical Theory and Practice. *Stat. Sci.* **2009**, *24*, 343–360.

Gee, G. W.; Bauder, J. W. Particle-size Analysis. In *Methods of Soil Analysis, Part 1: Physical and Mineralogical Methods*, 2nd ed; Klute, A., Ed.; Soil Science Society of America: Madison, 1986; pp 383–411.

Guillaume, T.; Muhammad, D.; Kuzyakov, Y. Losses of Soil Carbon by Converting Tropical Forest to Plantations: Erosion and Decomposition Estimated by $\delta 13C$. *Glob. Chang. Biol.* **2015**, *21*, 3548–3560. DOI: 10.1111/gcb.12907.

Hendarson, K. A.; Reis, M.; Blanco, C. C.; Pillar, V. D.; Printes, R. C.; Bauch, C. T.; Anand, M. Landowner Perceptions of the Value of Natural Forest and Natural Grassland in a Mosaic Ecosystem in Southern Brazil. *Sustainab. Sci.* **2016**, *11*, 321–330.

Hombegowda, H. C.; Van Straaten, O.; Kohler, M.; Holscher, D. On the Rebound: Soil Organic Carbon Stocks Can Bounce Back to Near Forest Levels when Agro-forests Replace Agriculture in Southern India. *Soil* **2016**, *2*, 13–23.

Houghton, R. A.; Goodale, C. L. Effects of Land-Use Change on the Carbon Balance of Terrestrial Ecosystems. In *Ecosystems and Land Use Change Defries*; R. S., Asner, G. P., Houghton, R. A., Eds.; American Geophysical Union: Washington, D. C., 2004.

Huang, M.; Ji, J.; Li, K.; Liu, Y.; Yang, F.; Tao, B. The Ecosystem Carbon Accumulation After Conversion of Grasslands to Pine Plantations in Subtropical Red Soil of South China. *Tellus* **2007**, *59B*, 439–448.

IPCC, Summary for Policymakers. In Climate Change 2007: The Physical Basis, Contribution of Working Group I to the Fourth Assessment Report of the Intergovernmental Panel on

Climate Change; Cambridge University Press: Cambridge, United Kingdom and New York, NY, USA, 2007.

IPCC, Core Writing Team. In Climate Change 2014: Synthesis Report; Contribution of Working Groups I, II and III to the Fifth Assessment Report of the Intergovernmental Panel on Climate Change; Pachauri, R. K., Meyer, L. A., Eds.; Geneva, Switzerland, 2014; p 151.

Iqbal, S.; Tiwari, S. C. Soil Organic Carbon Pool Under Different Land Uses in Achanakmar Amarkantak Biosphere Reserve of Chhattisgarh, India. *Curr. Sci.* **2016**, *110*, 771–73.

Jackson, R. B.; Banner, J. L.; Jobbagy, E. G.; Pockman, W. T.; Wall, D. H. Ecosystem Carbon Loss with Woody Plant Invasion of Grasslands. *Nature* **2002**, *418*, 623–6.

Jantke, K.; Muller, J.; Trapp, N.; Blanz, B. Is Climate-Smart Conservation Feasible in Europe? Spatial Relations of Protected Areas, Soil Carbon, and Land Values. *Environ. Sci. Policy* **2016**, *57*, 40–49.

Lal, R. Soil Carbon Sequestration Impacts on Global Climate Change and Food Security. *Science* **2004**, *304*, 1623–1627.

Lal, R. Forest Soils and Carbon Sequestration. *For. Ecol. Manag.* **2005**, *220*, 242–258.

Lal, R. Sequestration of Atmospheric CO_2 in Global Carbon Pools. *Energ. Environ. Sci.* **2008**, *1*, 86–100.

Lawrence, D.; Dodorico, P.; Delonge, M.; Diekmann, L.; Das, R.; Eaton, J. Ecological Feedbacks Following Deforestation Create the Potential for a Catastrophic Ecosystem Shift in Tropical Dry Forest. *PNAS* **2007**, *104*, 20696–20701.

Le Quere, C. *et al*. Global Carbon Budget 2016. *Earth Syst. Sci. Data* **2016**, *8*, 605–649.

Lenka, N. K.; Choudhury, P. R.; Sudhishri, S.; Dass, A.; Patnaik, U. S. Soil Aggregation, Carbon Build Up and Root Zone Soil Moisture in Degraded Sloping Lands Under Selected Agroforestry Based Rehabilitation Systems in Eastern India. *Agri. Ecosyst. Environ.* **2012**, *150*, 54–62.

Li, S.; Su, J.; Liu, W.; Lang, X.; Huang, X.; Jia, C.; Zhang, Z.; Tong, Q. Changes in Biomass Carbon and Soil Organic Carbon Stocks Following the Conversion from a Secondary Coniferous Forest to a Pine Plantation. *PLoS One* **2015**, *10*, 0135946.

Liao, C.; Luo, Y.; Fang, C.; Li, B. Ecosystem Carbon Stock Influenced by Plantation Practice: Implications for Planting Forests as a Measure of Climate Change Mitigation. *PLoS One* **2010**, *5*, e10867.

Majumdar, K.; Datta, B. K. Effects of Patch Size, Disturbances on Diversity and Structural Traits of Tropical Semi-Evergreen Forest in the Lowland Indo Burma Hotspot: Implication on Conservation of the Threatened Tree Species. *J. Mount. Sci.* **2016**, *13*, 1397–1410.

McGroddy, M. E.; Tanguy, D.; Lars, O. H. Scaling of C: N: P Stoichiometry in Forests Worldwide: Implications of Terrestrial Red Field-Type Ratios. *Ecology* **2004**, *85*, 2390–2401.

Mokany, K.; Rjohn, R.; Anatoly, S. P. Critical Analysis of Root: Shoot Ratios in Terrestrial Biomes. *Glob. Change Biol.* **2006**, *12*, 84–96.

Murty, D.; Kirschbaum, M. F.; McMurtrie, R. E.; McGilvray, H. Does Conversion of Forest to Agricultural Land Change Soil Carbon and Nitrogen? A Review of the Literature. *Glob. Change Biol.* **2002**, *8*, 105–123.

Murthy, I. K.; Gupta, M.; Tomar, S.; Munsi, M.; Tiwari, R. *et al.* Carbon Sequestration Potential of Agroforestry Systems in India. *J. Earth Sci. Climate Change* **2013**, *4*, 131.

Nandy, S.; Das, A. K. Comparing Tree Diversity and Population Structure Between a Traditional Agroforestry System and Natural Forests of Barak valley, Northeast India. *Inter. J. Biodiv. Sci. Ecosyst. Serv. Manag.* **2013**, *9*, 104–113.

Nath, A. J.; Brahma, B.; Lal, R.; Das, A. K. Soil and Jhum Cultivation. In *Encyclopedia of Soil Science,* 3rd ed.; CRC Press: Boca Raton, 2016.

Negash, M.; Starr, M. Biomass and Soil Carbon Stocks of Indigenous Agroforestry Systems on the South-Eastern Rift Valley Escarpment, Ethiopia. *Plant Soil* **2015**, *393*, 95–107.

Nelson, D. W.; Sommers, L. E. Total Carbon, Organic Carbon and Organic Matter. In *Methods of Soil Analysis, Part 2 Chemical and Microbiological Properties,* 2nd ed.; Page, A. L., Miller, R. H., Keeney, D. R., Eds.; The American Society of Agronomy and Soil Science Society of America, Madison: Wisconsin USA, 1982; pp 539–579.

Ngo, K. M.; Turner, B. L.; Muller-Landau, H. C.; Davis, S. J.; Larjavaara, M.; Hassan, N. F. B. N.; Lum, S. Carbon Stocks in Primary and Secondary Tropical Forests in Singapore. *For. Ecol. Manag.* **2013**, *296*, 81–89.

North Eastern Development Finance Corporation Ltd. *North Eastern Data Bank 2016*; NEDFi House: G. S. Road, Assam, India, 2016. http://databank.nedfi.com/content/assam (accessed Oct 15, 2017).

Ogle, S. M.; Breidt, F. J.; Paustian, K. Agricultural Management Impacts on Soil Organic Carbon Storage Under Moist and Dry Climatic Conditions of Temperate and Tropical Regions. *Biogeochem* **2005**, *72*, 87–121.

Oliveras, I.; Eynden, M. V. D.; Malhi, Y.; Cahuana, N.; Menro, C.; Zamora, F.; Haugaasen, T. Grass Allometry and Estimation of Above-Ground Biomass in Tropical Alpine Tussock Grasslands. *Austral. Ecol.* **2014**, *39*, 408–415.

Palm, C.; van Noordwijk, M.; Woomer, P.; et al. Carbon Losses and Sequestration After Land Use Change in the Humid Tropics. In *Slash-and-Burn Agriculture*; Palm, C.; Vosti, S.; Sanchez, P.; Ericksen, P., Eds.; Columbia University Press: New York, 2005; pp 41–63.

Pathak, K.; Nath, A. J.; Sileshi, G. W.; Lal, R.; Das, A. K. Annual Burning Enhances Biomass Production and Nutrient Cycling in Degraded *Imperata* Grasslands. *Land Degrad. Dev.* **2017**, *28*, 1763-1771.

Pereira, P.; Úbeda, X.; Mataix-Solera, J.; Oliva, M.; Novara, A. Short-term Changes in Soil Munsell Color Value, Organic Matter Content and Soil Water Repellency After a Spring Grassland Fire in Lithuania. *Solid Earth* **2014**, *5*, 209–225.

Poffenberger, M.; Barik, S. K.; Choudhury, D.; Darlong, V.; Gupta, V.; Palit, S.; Roy, I.; Singh, I.; Tiwari, B. K.; Upadhyay, S. *Forest Sector Review of Northeast India*; Community Forestry International; Santa Barbara, CA, USA, 2006.

Polasky, S.; Carpenter, S. R.; Folke, C.; Keeler, B. Decision-making Under Great Uncertainty: Environmental Management in an Era of Global Change. *Trends Ecol. Evol.* **2011**, *26*, 398–404.

Robertson, W. K.; Pope, P. E.; Tomlinson, R. T. Sampling Tool for Taking Undisturbed Soil Cores. *Soil Sci. Soc. Am. Proc.* **1974**, *38*, 855–857.

Roy, N.; Bezbaruah, M. P. *Agricultural Growth and Regional Economic Development (A Study of Barak Valley);* Mittal Publications: New Delhi, 2002.

Roy, P.; Kushwaha, S. P. S. Mirthy, M. S. R.; Roy, A.; Kushwaha, D.; Reddy, C. S.; Behera, M. D.; Mathur, V. B.; Padalia, H.; Saran, S.; Singh, S.; Jha, C. S.; Porwal, M. C. *Biodiversity Characterization at Landscape Level: National Assessment*; Indian Institute of Remote Sensing: Dehradun, 2012; p 140.

Saha, D.; Kukul, S. S.; Sharma, S. Landuse Impacts on SOC Fractions and Aggregate Stability in Typicustochrepts of Northwest India. *Plant Soil* **2011**, *339*, 457–470.

Sarkar, D. C.; Meitei, B.; Baisya, L. K.; Das, A.; Ghosh, S.; Khumlo, L. C.; Rajkhowa, D. Potential of Allow Chronosequence in Shifting Cultivation to Conserve Soil Organic Carbon in Northeast India. *Catena* **2015**, *135*, 321–327.

Schall, P.; Ammer, C. How to Quantify Forest Management Intensity in Central European Forests. *Eur. J. For. Res.* **2013**, *132*, 379–396.

Shapiro, S. S.; Wilk, M. B. An Analysis of Variance Test for Normality (Complete Samples). *Biometrika* **1965**, *52*, 591–611.

Sierra, C. A.; Del Valle, I. J.; Orrego, S. A.; Moreno, F. H.; Harmon, M. A.; Zapata, M.; Colorado, G. J.; Herrera, M. A.; Lara, W.; Restrepo, D. E.; Berrouet, L. M.; Loaiza, L. M.; Benjumea, J. F. Total Carbon Stocks in a Tropical Forest Landscape of the Porce Region, Columbia. *For. Ecol. Manag.* **2007**, *243*, 299–309.

Singh, S.; Dadhwal, V. K. *Manual on Spatial Assessment of Vegetation Carbon Pool of India*. Indian Institute of Remote Sensing (National Remote Sensing Centre), ISRO, Department of Space, Govt of India: Dehradun, 2009.

UNFCCC. Report of the Conference of the Parties on its Thirteenth Session, Held in Bali from 3 to 15 December 2007, Addendum Part Two: Action Taken by the Conference of the Parties at Its Thirteenth Session, 2008.

USDA. Keys to Soil Taxonomy. In *USDA Handbook*, 8th ed.; Soil Survey Staff: Washington, DC, 1998.

Vashum, K. T.; Jayakumar, S. Methods to Estimate Above-Ground Biomass and Carbon Stock in Natural Forests: A Review. *J. Ecosyst. Ecography* **2012**, *2*, 116.

Walkley, A.; Black, I. A. An examination of Degtjareff Method for Determining Soil Organic Matter and a Proposed Modification of the Chronic Acid Titration Method. *Soil Sci.* **1934**, *37*, 29–38.

Warren-Thomas, E.; Dolman, P. M.; Edwards, D. P. Increasing Demand for Natural Rubber Necessitates a Robust Sustainability Initiative to Mitigate Impacts on Tropical Biodiversity. *Conserv. Lett.* **2015**, *8*, 230–241.

Wei, X. R.; Shao, M. G.; Gale, W.; Li, L. H. Global Pattern of Soil Carbon Losses due to the Conversion of Forests to Agricultural Land. *Sci. Rep.* **2014**, *4*, 4062.

WMO (World Meteorological Organization). WMO Greenhouse Gas Bulletin: The State of Greenhouse Gases in the Atmosphere Based on Global Observations Through 2015, 2016. https://reliefweb.int/report/world/wmo-greenhouse-gas-bulletin-state-greenhouse-gases-atmosphere-based-global-observations (accessed Dec 7, 2016).

Yazaki, Y.; Shigeru, M.; Hiroshi, K. Carbon Dynamics and Budget in a *Miscanthus sinensis* Grassland in Japan. *Ecol. Res.* **2004**, *19*, 511–520.

Zar, J. H. *Biostatistical Analysis*, 4th ed.; Pearson Education India: New Delhi, India, 1999.

Zeigler, A. D.; Phelps, J.; Yuen, J. Q.; Webb, E. L.; Lawrence, D.; Fox, J. M.; Bruun, T. B.; Leisz, S. J.; Ryan, C. M.; Dressler, W.; Mertz, O.; Pascual, U.; Padoch, C.; Koh, L. P. Carbon Outcomes of Majorland-Cover Transitions in SE Asia: Great Uncertainties and REDD+ Policy Implications. *Glob. Chang. Biol.* **2012**, *18*, 3087–3099.

Zhang, M.; Huang, X.; Chuai, X.; Yang, H.; Lai, L.; Tan, J. Impact of Land Use Type Conversion on Carbon Storage in Terrestrial Ecosystems of China: A Spatial-Temporal Perspective. *Sci. Rep.* **2015**, *5*, 10233.

CHAPTER 6

Soil Carbon Sequestration

6.1 INTRODUCTION

Atmospheric carbon concentration has touched the fourth-century level in 2017 and is projected to increase day by day (Le Quere et al., 2016). Apart from other technological efforts, natural atmospheric carbon sinks are very significant in the removal of carbon from the atmosphere. Carbon sequestration and maintaining it in an ecosystem level has been highly appreciated to keep the atmospheric carbon concentration up to a considerable level to minimize the risks of climate change (Sarkar et al., 2015). Vegetative and soil pools are two interlinked and significant pools that maintain ecosystem carbon budget (Le Quere et al., 2016). Direct accumulation of atmospheric carbon in vegetative parts of green plants during the process of photosynthesis is the major carbon stock of tropical forests, accounts about two-third of the total ecosystem carbon (Ngo et al., 2013). The translocated form of vegetation, dead microorganisms, and other dead animals constitute the organic carbon stock in soil (Nelson and Sommers, 1996). Upper soil layers usually possess higher organic matter with higher micro-organic activity than that of the lower soil depths. Therefore, out of total organic carbon (1550 Gt) stored in the soil (Le Quere et al., 2016; Lal, 2016), a large proportion (216 Gt) is confined in the first 1 m soil depth under forests (IPCC, 2007). However, the rate of soil carbon sequestration in terrestrial ecosystems is greatly influenced by its vegetative cover (Brahma et al. 2017). Furthermore, vegetative cover changes with environmental gradients, human interventions, and land use change (Burke et al., 1989). Thus, the land use change and human intervention have direct influence on the soil carbon stock (FAO, 2010). Land use change from forests into croplands or other managed agro-ecosystems reduces 19–40% of stored SOC (Guillaume et al., 2015; De Straaten et al., 2015; De Blecourt et al., 2013). In addition, land use change through deforestation and forest degradation acts as a C source (Van Der Werf et al., 2009). Furthermore, it has been estimated that deforestation and forest degradation of tropical forests contribute 12%–15% of global anthropogenic carbon dioxide (CO_2) emission (Le Quere et al., 2016). Therefore, changes in soil organic carbon (SOC) stock in forest soil due to land use change may induce a large effect in atmospheric carbon concentration (Harris et al., 2012; Don et al., 2011).

Rubber tree (*Hevea brasiliensis*) is natural to Brazilian forests and was introduced into Asia for commercial latex production in 1876 (ICAR, 2003). Due to increasing commercial value and demands of natural rubber, the cultivation of rubber trees has spanned over the Asia, Africa, and America; and covering currently ~10 million ha (M ha) area (Statista, 2016). Since the humid and sub-humid tropical climates are suitable for rubber cultivation, the overwhelming growth of rubber cultivation by replacing tropical forests has been observed during last two decades (Fox and Castella, 2013; Bahamondez et al., 2010; Priyadarshan et al., 2005). The estimated increasing trend of commercial rubber cultivation around the world is 0.10 M ha per annum during the period of 2005–2010 (Statista, 2016). In the face of rapid expansion of rubber plantation, biodiversity conservation, and provisioning ecosystem services are under serious threat (He and Martin, 2015). However, with this negative impact, the role of rubber plantation is still in controversy with positive atmospheric carbon sink capacity in soil (Brahma et al., 2016; Li et al., 2012). Therefore, the role of rubber plantations in atmospheric carbon dynamics is a crucial need for developing the future policies. In the present study, four different aged rubber plantations (6, 15, 27, and 34 year old) were selected to study whether or not SOC stock under rubber plantation can attain equilibrium with that under natural forest after 34 years of plantation development.

6.2 STUDY SITE AND STAND CHARACTERISTICS

The study area is located in the Barak Valley of Assam in northeast India (NEI). It falls within the range of the Himalayan foothills and the Barak River Basin. The Barak Valley consists of three districts of Assam (Cachar, Hailakandi, and Karimganj) covering an area of ~7000 km^2 (NEDFI, 2016). The valley is characterized by tropical humid climate, and experiences mean annual precipitation of 3500 mm, temperature range of

13–37°C and relative humidity of 93.5% (NEDFI, 2016). The dominant soil is classified as the Barak series, which is fine, mixed, hyperthermic family of *Aeric Endoaquepts* (USDA, 1998).

Natural forest from Karimganj district was selected for the present study (24°40′N, 092°46′E). The forest is classified as tropical wet evergreen forest (Champion and Seth, 2005), and dominant tree species are *Cynometra polyandra, Ficus hispida, Litsea* sp., *Lindera* sp., and *Oroxylum indicum*, etc. The grasslands in Barak Valley originally developed as result of degradation of natural forests and were subsequently invaded by *Imperata cylindrica*. Annual harvesting of above ground part of the *I. cylindrica* for thatching and roofing purposes is very common in NEI (Pathak et al., 2017). Burning of above ground residues is also a common managerial system practiced annually to maintain the soil nutrients under *I. cylindrica* grasslands (Pathak et al., 2017). Rubber plantation in Barak Valley is a result of consecutive land use change from natural forest to grassland and to rubber plantations. Rubber plantations were established for production of natural rubber latex, which is an important source of monitory income in the region. Clearing of natural forests or grasslands prior to the development of rubber plantation is a mere practice. Accumulation of leaf litter around the rubber tree to maintain the soil moisture is a common practice. For the present study, rubber plantations of different ages situated at 200–300 m from each other were chosen from the Karimganj district (24°36′ N, 092°23′ E). The age of the rubber tree plantations and the history of the land use were confirmed from the forest's official records and interview of local people.

6.3 SOIL SAMPLING AND ANALYSES

Soil samples from differently aged plantations and natural forest were collected during February and March, 2015. Four plots with 25 m × 25 m were placed randomly in each study site and three pits (1 m width × 1 m length × 1 m depth) were dug in each plot. Thus, a total of 12 pits were dug within each plantation and also in natural forest. From each soil pit soil samples were collected with soil core (5.6 cm inner diameter) from 0 to 20 cm, 20 to 50 cm, and 50 to 100 cm depth. Three soil samples from each depth were collected by inserting the core horizontally up to 10 cm and transported into the laboratory. Thus, for each study site, a total of 36 composite soils were prepared for subsequent air drying. Air dried soils were sieved through 100-micron mesh for SOC concentration analysis. The SOC concentration (%) of each of the soil sample was determined through the wet oxidation method (Jackson, 1958).

For soil bulk density (BD) estimation, two additional samples were collected from each depth using the same soil corer from each soil pit. Thus, a total of 72 soil samples were collected from each study site. Soil BD was calculated after oven drying the soil samples at 105°C (Robertson et al., 1974). The average value of two soil samples for each depth of the same soil pit was considered as the dry BD for that specific depth. The average value for three consecutive soil depth of a soil pit was considered as the dry BD for 0–100 cm soil depth.

The following formula was used for calculating the SOC stock (Mg ha^{-1}; Blanco-Canqui and Lal, 2008):

$$\text{SOC stock}\left(\text{Mg ha}^{-1}\right) = 10^4 \left(\frac{\text{m}^2}{\text{ha}}\right) \times \text{soil depth}(\text{m}) \times \text{BD}\left(\frac{\text{Mg}}{\text{m}^3}\right) \times \frac{\text{SOC}(\%)}{100},$$

where BD is bulk density and SOC is soil organic carbon concentration.

The SOC stock for 0–20, 20–50, and 50–100 cm soil has been determined by multiplying the SOC concentration with BD of the respective soil depth. The SOC stock for 0–100 cm depth was estimated by summing up the SOC stock values for three consecutive soil depths. The SOC stock for a specific soil depth under a specific plantation age/natural forest was calculated from the average SOC stock (Mg ha^{-1}) values of all the soil pits. The SOC sequestration rate for different aged plantation was calculated from SOC stock difference between any two plantation ages divided by differences in age of the plantation.

6.4 STATISTICAL ANALYSIS

The significant differences in BD, SOC concentrations, and SOC stocks of different soil depths among the study sites were enquired by applying one-way ANOVA at 95% confidence level (Zar, 1999). Thereafter, if the ANOVA indicated any significant differences ($P < 0.05$), Tukey honestly significant difference (HSD) post-hoc test was applied (Pereire et al., 2014). All statistical calculations to prepare this chapter were done using MS-Excel 2003 and IBM SPSS Statistics 21 software.

6.5 SOIL BULK DENSITY UNDER RUBBER PLANTATIONS

The observed BD of different soil depths under rubber tree plantations ranged from 1.14 to 1.29, 1.19 to 1.45, and 1.40 to 1.56 Mg m^{-3} for 0–20, 20–50, and 50–100 cm depth, respectively, whereas under natural forest, it ranged from 1.20 to1.38 for 0–20 and 50–100 g cm^{-3} soil depth, respectively (Fig. 6.1).

The upper layer of soil contained higher organic matter with higher microorganic activities that lead to lower BD of the surface soil than that of sub-surface soils (Bot and Benites, 2017). Similar results have been observed in the present study under different ages of rubber tree plantations although the BD values were not significantly different among the soil depths ($P > 0.05$). The BD of the soil is also affected by the land use pattern and anthropogenic activities (Pizzeghello et al., 2017). However, there was no effect of land use change on BD values for a specific soil depth ($P > 0.05$).

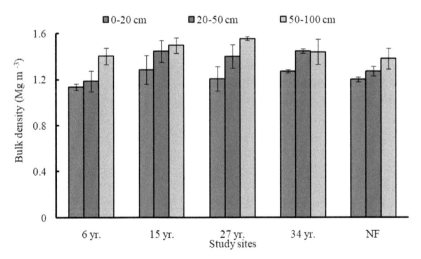

FIGURE 6.1 **(See color insert.)** Soil bulk density (Mg m^{-3}) for different soil depths under different aged rubber plantations and natural forest. Error bar represents standard error of the mean. In the figure, 6, 15, 27, and 34 years are rubber tree plantations under different ages and NF represents the natural forest.

6.6 EFFECT OF LAND USE CHANGES IN SOIL ORGANIC CARBON CONCENTRATION

The SOC concentration (g kg^{-1}) for different soil depths under different aged rubber tree plantations are presented in Figure 6.2. The estimated SOC concentration for 0–20, 20–50, and 50–100 cm soil depths under different aged rubber tree plantations ranged from 1.30 to 1.77, 0.82 to 1.08, and 0.46 to 0.71 g kg^{-1} (Fig. 6.2). However, under natural forest, the SOC concentration for the different soil depths ranged from 1.26 to 1.65 g kg^{-1} (Fig. 6.2).

SOC concentration is the fraction of organic carbon present in the soil. Mostly carbon of the plant litters and other residues enters into the soil through decomposition. Therefore, the SOC concentration directly depends on different vegetation types (Zhang et al., 2017; Guo et al., 2016). Moreover, other physical and biological factors like sunlight availability, precipitation, and moisture content of the soil influence the SOC content (Tulina et al., 2009; Sun et al., 2016). The SOC concentration also varies with the depth of the soil. The surface soil contains higher SOC concentration than sub-surface soil (Gupta et al., 2009), a similar trend has been observed in the present study. The SOC concentration for the upper soil depth (0–20 cm) under each of the study site was observed to be significantly higher ($P < 0.05$) than that of the 20–50 cm and 50–100 cm depth, respectively. Such similar trend of SOC concentration with depth was reported from rubber plantations in Brazil (Maggiotto et al., 2014). However, SOC concentration of the soil is highly controlled by the land use pattern, management activities (Pizzeghello et al., 2017). In addition to that, in plantation ecosystem, fresh organic matter input through litter varies with the age of the plantation (Wang et al., 2016), which leads to the higher concentration of SOC under higher aged plantations than that of under younger age (Kalita et al., 2016). Therefore, in the present study a significant positive trend of SOC concentration with age of the plantation has been observed for 0–20 cm ($F = 63.14$, $P = 0.00$) and 20–50 cm ($F = 193.10$, $P = 0.00$) soil depth. A little higher SOC concentration for 0–20 cm soil depth under 6-year-old plantation than that of 15-year-old plantation was observed. However, this variation was not significant when Tukey's HSD test was applied ($P = 0.91$). The higher SOC concentration in 6-year-old plantation is attributed to manuring practices commonly adopted during the initial stage of rubber tree plantations (The Rubber

Board, 2011). A significant ($F = 398.52$, $P = 0.00$) decreasing trend in SOC concentration for 50–100 cm soil depth has been observed with the age. No significant difference for 50–100 cm soil depth has been observed between 15- and 34-year-old rubber tree plantations ($P = 0.13$). The estimated SOC concentration for 0–20 cm soil depth for natural forest was found to be lower than 27 and 34 year rubber stand; whereas, the estimated SOC concentration for 20–50 and 50–100 cm soil depth was higher than that under rubber plantations.

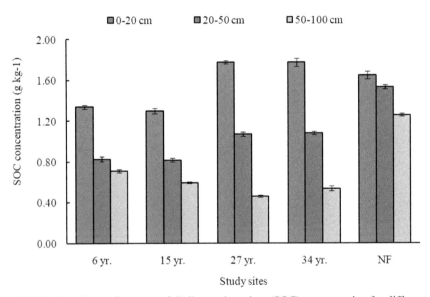

FIGURE 6.2 (See color insert.) Soil organic carbon (SOC) concentration for different soil depth under different aged rubber plantations and natural forest. Error bar represents standard error of the mean. In the figure, 6, 15, 27, and 34 years are rubber tree plantations under different ages and NF represents the natural forest.

6.7 LAND USE CHANGE AND SOIL CARBON STOCK

The estimated SOC stock for 0–100 cm soil depth ranged from 109.26 to 130.24 Mg ha^{-1} for 6- and 34-year-old plantations, respectively (Fig. 6.3). The present study shows the SOC stock under natural forest was 184.83 Mg ha^{-1} (Fig. 6.3).

SOC stocks are positively correlated with the SOC concentration and BD. The SOC concentrations are usually negatively correlated with the BD of the soil. Moreover, land use change affects the SOC stocks. Soils under natural forest store higher SOC stock than that of under plantation and agroforestry systems (Brahma et al., 2017). Rubber plantations in NEI are usually the converted form from forests or grassland. Therefore, carbon stored under natural forest or grassland may be lost during the process of rubber plantation development. Present study shows SOC stock under rubber plantations increased with an increase in age of the plantation and soil depth. The SOC stock for the upper soil depth (0–20 cm) of the rubber tree plantations increased from 30.43 Mg ha^{-1} under 6 year to 44.99 Mg ha^{-1} under 34 year plantation. Soils under rubber plantations of Brazil stores 52.42–101.45 Mg ha^{-1} of SOC up to the 60 cm depth under 2–26-year-old plantations (Maggiotto et al., 2014; Wauters et al., 2008). In the present study, a comparable range of SOC stock with 59.69–91.94 Mg ha^{-1} up to the 50 cm soil depth was estimated under 6–34 year rubber plantation. Natural forest soils up to 50 cm of the present study stores 10% more SOC than that under rubber plantations. The SOC stock in lower depth (50–100 cm) shows the decreasing trend with age of the plantation (49.57–35.81 Mg ha^{-1} for 6 and 27 years, respectively). Lower depth of the soils is restricted from the availability of litter productions. A little increase in SOC stock in lower depth after an age of 27 years of the rubber plantation was observed. This increment may be subjected to the higher fine root production under mature plants and percolation of soil carbon from the upper soils to the lower by this time period. Natural forests sink more SOC than human-induced plantations or croplands (Lal, 2008). Therefore, in the present study SOC stock up to 100 cm soil depth under natural forest was estimated significantly ($P < 0.05$) higher than that under rubber plantations. Even after 34 years of rubber plantation development, it contained 30% less SOC than the natural forest. Conversion of tropical forests into croplands like oil palm (*Elaeis guineensis*), rubber and cacao (*Theobroma cacao*) resulted in loss of SOC by 50% (De Straaten et al., 2015), which is comparable to the present findings. A progressive increment of 4%, 9%, and 5% of SOC stock was observed for 6–15 years, 15–27 years, and 27–34 years of rubber plantations, respectively. However, Tukey's HSD test revealed that SOC stock for 0–100 cm soil depth under 34-year-old plantations is significantly ($P = 0.04$) increased from 6-year-old plantation and the estimated increment was 19% with C sequestration rate of 0.75 Mg ha^{-1} year^{-1}.

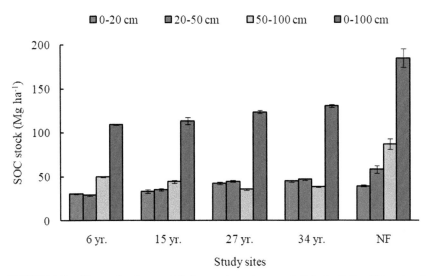

FIGURE 6.3 **(See color insert.)** Soil organic carbon stock (Mg ha^{-1}) for different soil depths under different aged rubber plantations and natural forest. Error bar represents standard error of the mean. In the figure, 6, 15, 27, and 34 years are rubber tree plantations under different ages and NF represents the natural forest.

6.8 SOIL CARBON SEQUESTRATION UNDER RUBBER PLANTATIONS

SOC sequestration rate under rubber plantations increased with the age of the stand and it ranged from 0.18 to 0.99 Mg ha^{-1} year^{-1} (Fig. 6.4).

Mandal and Islam (2010) reported that the rate of carbon sequestration increased quadratically which accounted for 1.071 Mg ha^{-1} year^{-1} under the 5-year-old plantation, 0.930 Mg ha^{-1} year^{-1} under the 10-year-old plantation, and 0.729 Mg ha^{-1} year^{-1} under the 20-year-old plantation. Research result from NEI suggested that degraded soils under managed rubber plantations have considerable potential for carbon sequestration (Brahma et al., 2017). Lal et al. (1995) estimated that with resilient land management practices, degraded soils can sequester 0.10–1.0 Mg ha^{-1} year^{-1}. Cheng et al. (2007) reported that degraded soils under rubber plantations have

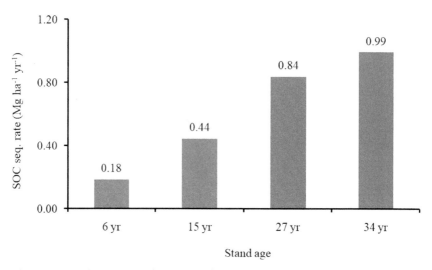

FIGURE 6.4 SOC sequestration rate or time-averaged SOC sink for different ages of rubber plantations.

more C sequestration potential than by primary and secondary rain forests. However, soil C sequestration eventually achieves an equilibrium point and the ability of the soil to sequester C decreases over time as previously depleted stocks are replenished and the soil returns to an equilibrium condition. However, the time it takes for soils to return to equilibrium is highly variable (Mandal and Islam, 2010).

6.9 CONCLUSIONS

In conclusion, following points emerged out: (1) with increase in age of the plantation SOC stock increased by 19% from 6- to 34-year-old plantation, (2) although increase in SOC stock with age was consistent it was 30% less than that under natural forest, and (3) with the C sequestration rate of 0.99 Mg ha^{-1} year^{-1}, it is projected that rubber plantations require 90 years to attain equilibrium in SOC stock with that under natural forest.

KEYWORDS

- plantation age
- sequestration rate
- land use conversion
- forest degradation
- SOC stock

REFERENCES

Bahamondez, C. C.; Csoka, T.; Drichi, P.; Filipchuk, P.; De Jong, A.; Alvarez, B. H. *Global Forest Resources Assessment 2010 Main Report*; FAO. 2010.

Blanco-Canqui, H.; Lal, R. No-tillage and Soil-profile Carbon Sequestration: An On-Farm Assessment. *Soil Sci. Soc. Am. J.* **2008,** *72*, 693–701.

Bot, A.; Benites, J. *The Importance of Soil Organic Matter: Key to Drought-Resistant Soil and Sustained Food and Production*; FAO: Rome, 2017; pp 11–14.

Brahma, B.; Nath, A. J.; Das, A. K. Managing Rubber Plantations for Advancing Climate Change Mitigation Strategy. *Curr. Sci.* **2016,** *110*, 2015–2019.

Brahma, B.; Sileshi, G. W.; Nath, A. J.; Das, A. K. Development and Evaluation of Robust Tree Biomass Equations for Rubber Tree (*Hevea brasiliensis*) Plantations in India. *For. Ecosyst.* **2017,** *4,* 14–24.

Champion, H. G.; Seth, S. K. *A Revised Survey of the Forest Types of India*; Natraj Publishers: Dehradun, 1968 (Reprinted 2005); p 404.

De Blecourt, M.; Brumme, R.; Xu, J.; Corre, M. D.; Veldkamp, E. Soil Carbon Stocks Decrease Following Conversion of Secondary Forests to Rubber (*Hevea brasiliensis*) Plantations. *PLoS One* **2013,** *8,* 69357.

Don, A.; Schumacher, J.; Freibauer, A. Impact of Tropical Land-Use Change on Soil Organic Carbon Stocks—A Meta-Analysis. *Glob. Change Biol.* **2011,** *17*, 1658–1670.

FAO. *Global Forest Resource Assessment*; Food and Agricultural Organization of The United Nations (FAO): Rome, 2010.

Fox, J.; Castella, J. C. Expansion of Rubber (*Hevea brasiliensis*) in Mainland Southeast Asia: What Are the Prospects for Smallholders? *J. Peasant Stud.* **2013,** *40*, 155–170.

Guillaume, T.; Muhammad, D.; Kuzyakov, Y. Losses of Soil Carbon by Converting Tropical Forest to Plantations: Erosion and Decomposition Estimated by δ13C. *Glob. Change Biol.* **2015**.

Gupta, N.; Kukal, S. S.; Bawa, S. S.; Dhaliwal, G. S. Soil Organic Carbon and Aggregation Under Poplar Based Agroforestry System in Relation to Tree Age and Soil Type. *Agrofor. Syst.* **2009,** *76,* 27–35.

Harris, N. L.; Brown, S.; Hagen, S. C.; Saatchi, S. S.; Petrova, S.; Salas, W.; Hansen, M. C.; Potapov, P. V.; Lotsch, A. Baseline Map of Carbon Emissions From Deforestation in Tropical Regions. *Science* **2012**, *336*, 1573–1576.

He, P.; Martin, K. Effects of Rubber Cultivation on Biodiversity in the Mekong Region. CAB Reviews 10, 2015. DOI: 10.1079/PAVSNNR201510044.

ICAR. *Handbook of Agriculture. Indian Council of Agriculture*; Pusa: New Delhi, 2003.

IPCC. Summary for Policymakers. In *Climate Change 2007: The Physical Basis, Contribution of Working Group I to the Fourth Assessment Report of the Intergovernmental Panel on Climate Change*; Cambridge University Press: Cambridge, United Kingdom and New York, NY, USA, 2007.

Jackson, M. L. *Soil Chemical Analysis*; Prentice Hall, Englewood, Cliffs: New Jersey, 1958.

Kalita, R. M.; Das, A. K.; Nath, A. J. Assessment of Soil Organic Carbon Stock Under Tea Agroforestry System in Barak Valley, North East India. *Inter. J. Ecol. Environ. Sci.* **2016**, *42* (2), 175–182.

Lal, R. Sequestration of Atmospheric CO_2 in Global Carbon Pools. *Energ. Environ. Sci.* **2008**, *1* (1), 86–100.

Lal, R. Soil Health and Carbon Management. *Food Ener. Secur.* **2016**, *5*, 212–222.

Le Quéré, C.; Andrew, R. M.; Canadell, J. G.; Sitch, S.; Korsbakken, J. I.; Peters, G. P. et al. Global Carbon Budget 2016. *Earth Syst. Sci. Data* **2016**, *8*, 605.

Li, H.; Ma, Y.; Liu, W.; Liu, W. Soil Changes Induced by Rubber and Tea Plantation Establishment: Comparison with Tropical Rain Forest Soil in Xishuangbanna, SW China. *Environ. Manag.* **2012**, *50* (5), 837–848.

Maggiotto, S. R.; Oliveira, D.; Marur, C. J.; Stivari, S. M. S.; Leclerc, M.; Wagner-Riddle, C. Potential Carbon Sequestration in Rubber Tree Plantations in the Northwestern Region of the Paraná State, Brazil. *Acta Scientiarum. Agro.* **2014**, *36*, 239–245.

Nelson, D. W.; Sommers, L. E. Total Carbon, Organic Carbon and Organic Matter. In *Methods of Soil Analysis. Part 2 Chemical and Microbiological Properties;* 2nd ed., Page, A. L., Miller, R. H., Keeney, D. R., Eds.; The American Society of Agronomy and Soil Science Society of America: Madison, Wisconsin USA, 1982, pp 539–579.

Ngo, K. M. Carbon Stocks in Primary and Secondary Tropical Forests in Singapore. *For. Ecol. Manag.* **2013**, *296*, 81–89.

North Eastern Development Finance Corporation Ltd. North Eastern Data Bank 2016. NEDFi House, G. S. Road, Assam, India, 2016. http://databank.nedfi.com/content/assam (accessed Oct 15, 2017).

Pathak, K.; Nath, A. J.; Sileshi, G. W.; Lal, R.; Das, A. K. Annual Burning Enhances Biomass Production and Nutrient Cycling in Degraded *Imperata* Grasslands. *Land Degrad. Dev.* **2017**.

Pereira, P.; Úbeda, X.; Mataix-Solera, J.; Oliva, M.; Novara, A. Short-term Changes in Soil Munsell Colour Value, Organic Matter Content and Soil Water Repellency After a Spring Grassland Fire in Lithuania. *Solid Earth* **2014**, *5*, 209–225.

Pizzeghello, D.; Francioso, O.; Concheri, G.; Muscolo, A.; Nardi, S. Land Use Affects the Soil C Sequestration in Alpine Environment, NE Italy. *Forests* **2017**, *8* (6), 197.

Priyadarshan, P. M.; Hoa, T. T. T.; Huasun, H.; De Goncalves, P. S. Yielding Potential of Rubber (*Hevea brasiliensis*) in Sub-optimal Environments. *J. Crop Impro.* **2005**, *14*, 221–247.

Robertson, W. K.; Pope, P. E.; Tomlinson, R. T. Sampling Tool for Taking Undisturbed Soil Cores. *Soil Sci. Soc. Am. Proc.* **1974**, *38*, 855–857.

Sarkar, D. C.; Meitei, B.; Baisya, L. K.; Das, A.; Ghosh, S.; Khumlo, L. C.; Rajkhowa, D. Potential of Allow Chronosequence in Shifting Cultivation to Conserve Soil Organic Carbon in Northeast India. *Catena* **2015**, *135*, 321–327.

Statista, The Statistics Portal, 2016. https://www.statista.com/statistics/238900/rubber-plantations-areas-worldwide/ (accessed Sept 20, 2016)

Straaten De, O. V.; Corre, M. D.; Wolf, K.; Tchienkoua, M.; Cuellar, E.; Matthhews, R. B.; Veldkamp, E. Conversion of Lowland Tropical Forests to Tree Cash Crop Plantations Loses Up to One-Half of Stored Soil Organic Carbon. *PNAS* **2015**, *112*, 9956–9960.

The Rubber Board. *Rubber Growers Guide 2011*; Government of India: Kottayam, Kerala, India, 2011.

USDA. Keys to Soil Taxonomy. In: *USDA Handbook*, 8th ed.; Soil Survey Staff: Washington, DC, 1998.

Van Der Werf, G. R.; Morton, D. C.; DeFries, R. S.; Olivier, J. G. J.; Kasibhata, P. S.; Jachson, R. B.; Collatz, G. J.; Randerson, J. T. CO_2 Emissions from Forest Loss. *Nat. Geosci.* **2009**, *2*, 737–738.

Wang, G. B.; Deng, F. F.; Xu, W. H.; Chen, H. Y. H.; Ruan, H. H. Poplar Plantations in Coastal China: Towards the Identification of the Best Rotation Age for Optimal Soil Carbon Sequestration. *Soil Use Manag.* **2016**, *32*, 303–310.

Wauters, J. B.; Coudert, S.; Grallien, E.; Jonard, M.; Ponette, Q. Carbon Stock in Rubber Tree Plantations in Western Ghana and Matogrosso (Brazil). *For. Ecol. Manag.* **2008**, *255*, 2347–2361.

Zar, J. H. *Biostatistical Analysis,* 4th ed.; Pearson Education India: New Delhi, India, 1999.

Zhang, M.; Fu, X. H.; Feng, W. T.; Zou, X. Soil Organic Carbon in Pure Rubber and Tea-Rubber Plantations in South-Western China. *Trop. Ecol.* **2007**, *48*, 201–207.

CHAPTER 7

Fine Root Dynamics and Carbon Management

7.1 INTRODUCTION

Carbon as carbon dioxide (CO_2) gas in the atmosphere is characterized to trap the earth's emitting long wave radiations and eventually keeps the earth warm. It is one of the most important greenhouse gases maintaining the earth's heat budget and to maintain it, the concentration of atmospheric CO_2 is crucial. Global average atmospheric carbon concentration is increasing day by day and has reached its ever highest level in 2016 at 402.9 ppm. The inconsistency of atmospheric CO_2 concentration is a natural phenomenon but eventually seems to be increasing year to year. Based on the ice core data it was observed that 800,000 years ago the concentration was below 200 ppm and around 330,000 years ago it reached the highest level of around 300 ppm (Lüthi et al., 2008). It took 470,000 years to increase the atmospheric carbon concentration by >103 ppm. Whereas, a further increment of 100 ppm atmospheric CO_2 took place in only 330,000 years. Natural sources like decomposition, oceanic release, respiration, and other natural phenomenon were mostly responsible for this increment during historical period. However, during the pre-industrial period in 1750, the CO_2 concentration was observed to be 277 ppm, lower than that of the highest level of historical period (Joos and Spahni, 2008). Considering 1750 as a baseline, the increment of 105 ppm in atmospheric CO_2 concentration was perturbed only within recent 265 years. In addition to the natural sources of CO_2 emissions, numerous anthropogenic sources were being credited for this dramatic increment of CO_2 in the recent centuries. During the initial stage of industrialization, the chief source of carbon to the atmosphere was recognized as deforestation and other land use changes, whereas fossil fuel burning as a chief source of atmospheric carbon was focused after 1920 (Ciais et al., 2013). In order to fulfill the demand of increasing world population fossil fuel burning increases in a dramatic way and produces around 21.3 Gt of CO_2 yearly. The estimated CO_2 emission from fossil fuel burning and industrial activities has been reached equivalent to 9.3 Gt carbon per year from 2006 to 2015 (Le Quere et al., 2016). The estimated C emission from the land use changes related activities like deforestation, afforestation, shifting cultivation, cropland harvest and peat fires over the same decade was around 1Gt carbon per year (Le Quere et al., 2016). If the current trend of C emission continue, the projected carbon concentration will reach within the range of 463–623 ppm by 2050 accompanied with 0.8–2.6°C atmospheric temperature increase with respect to 1990 (IPCC). The consequences of this increased temperature will be loss of numerous living species with disappearance of huge land masses. In order to refrain these projections to take place, reducing the emission through anthropogenic activities and increasing carbon removal from the atmosphere are of most interests. Therefore, understanding and management of terrestrial carbon stock has become a highly appreciated mechanism to mitigate the adverse climate change effects on the earth (UNFCCC, 2008). By contributing ~46% of the global carbon balance, natural and plantation forests of tropical region play a significant role in maintaining atmospheric carbon concentration (Wei et al., 2014; Bloom et al., 2016).

Rubber tree (*Hevea brasiliensis*) is one of the most dominating tree crops around the tropical belt covering around ~10 Mha land area. Apart from its economic profitability, rubber plantations have been preferred for accelerated carbon sequestration (Brahma et al., 2016). The ecosystem carbon sink capacity of rubber tree plantations are comparable to the traditionally managed system and natural forests and higher than that of under degraded forest, *Areca catechu* plantation and grasslands of the same region of northeast India (Brahma et al., 2017). Biomass and soil are the two important pools that constitute the major terrestrial carbon stock. However, the biomass and atmospheric derivative carbon stored into the soil are much more than that of under biomass and atmosphere (Lal, 2004). Translocation of biomass carbon into the soil chiefly depends upon the biomass allocation in aboveground and belowground vegetative part. Belowground biomass carbon in the form of fine root acts as the chief stream in the soil carbon and an essential section of global carbon cycle (Graaff et al., 2013; Kong and Six, 2010; Rasse et al., 2005; Puget and Drinkwater, 2001). Globally about 33% of the annual net primary production of terrestrial ecosystems has been translocated into the soil through fine root dynamics (Jackson et al.,

1997). Fine roots (<2 mm diameter) are very dynamic in turnover and are also merely responsible for uptake and transportation of water and nutrient (Gower et al., 2014; Olmo et al., 2014; Finer et al., 2007). The production and turnover rate of fine roots varies with the spatial and temporal variations with the additional variations of species diversity and climate (Olmo et al., 2014; Al Afas et al., 2008; Hendricks et al., 2006; Leuschner et al., 2004; Vagt et al., 1996). Therefore, study on fine root dynamics for the dominant plantations of any region can advance our understanding on impact of fine roots on carbon sink management. The contribution of belowground biomass in carbon management under rubber tree plantations has been little studied (Brahma et al., 2017). Therefore, in the present chapter fine root production (FRP) and their mean residence time under chronosequence of rubber plantations has been explored. Relation of these traits with stand age and rainfall was also explored in the present study.

7.2 METHODS FOR FINE ROOT BIOMASS STUDY

Fine roots of the surface soil layer are very dynamic in nature and play a significant role in the soil carbon flux (Chimento and Amaducci, 2015). Therefore, to estimate the fine root biomass under chronosequence of rubber plantations, four different age plantation namely, 6-, 15-, 27-, and 36-year-old stands were selected from Barak Valley part of northeast India. Soil samples were collected from 0 to 20 cm depth of the soil with a sharp corner of 5.6 cm inner diameter from December, 2014 to November, 2015. For fine root collection, three random plots with 10 m × 10 m area were laid within each of the representative plots (250 m × 250 m) under each rubber stand. Four representative plots with 250 m × 250 m area were previously selected under each rubber stand. Therefore, a total of 12 plots (10 m × 10 m) were laid down to collect core samples and at each of the monthly sampling date, two random samples were collected from each of the 12 plots. Thus, a total of 24 core samples were collected for monthly root biomass estimation from each rubber stand and taken to the laboratory for further analysis. Data for each 10 m × 10 m plot was primarily averaged to observe the variations among the plots. Thereafter, values of 12 plots were averaged to see the monthly variations and that were used to estimate the fine root biomass of the stand.

In the laboratory, soil samples were soaked overnight and roots were cleaned from soil and organic residues by a gentle stream of water over a fine sieve. Roots less than or equal to 2 mm diameter were separated manually (Fig. 7.1). Live and dead roots were also carefully separated soon after removing the roots of other plants. Live roots of rubber plants were directly observed as light in color, resilient to break and decay free, whereas the dead roots were brown or black in color and easily broken (Jessy et al., 2013; Brassard et al., 2011; Bennett et al., 2002). Fine roots of other herbs and shrubs were separated based on color, branching pattern, and texture (Brassard et al., 2011; Bennett et al., 2002; Thaler, 1996). For estimating fine root biomass, dry weight of the fine roots was taken after drying at 70°C for 72 h (Jamro et al., 2015). From the sequential coring data, the estimated annual fine root biomass (RB) for live and dead root were separately calculated for each plot by following the maximum–minimum method (McClaugherty et al., 1982) and averaged to report as the estimated yearly biomass (Mg ha^{-1}). For a specific plot the minimum biomass (RB_{min}) over the period was subtracted from the maximum biomass (RB_{max}) recorded for the same plot observed over the course of the sampling period (Lemann and Zech, 1998; Ruess et al., 1996; Publicover and Vogt, 1993; Aber et al., 1985; McClaugherty et al., 1982). The followed expression was:

$$ERB = RB_{max} - RB_{min}.$$

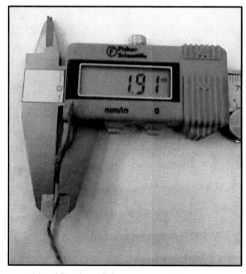

FIGURE 7.1 Fine root identification and measurement.

Fine Root Dynamics and Carbon Management

FIGURE 7.2 Soil collected for fine root has been kept for overnight water soaking.

7.2.1 Fine Root Decomposition

Fine root decomposition rate or decay constant k (year^{-1}) under different ages of rubber stands has been determined through mesh bag technique (Fahey et al., 1988; McClaughertyet al., 1984). Fine roots were collected from the topsoil layer (0–20 cm) of each rubber stand and washed carefully. Thereafter, the washed roots were air dried for one week. A total of 150 mesh bags of 6 cm × 6 cm size and 0.3 mm × 0.3 mm mesh size (Jamro et al., 2015) with 2 g of air-dried root material was placed in three different places under each rubber stand in December, 2014 (Fig. 7.2). At each monthly sampling, three litter bags were collected from each of the three different places. Thus, nine mesh bags were collected at monthly interval from each rubber stand (Fig. 7.3) and continued for 1 year. Therefore, at each sampling date, a total of 36 numbers of mesh bags were collectively retrieved from all the rubber stands. During the laboratory analysis after collecting the mesh bags from the field the residual roots were carefully removed from the bag and washed gently over the mesh to remove the adherent soils. Soon after washing the residual roots were dried at 65°C for 48 h and took the dry weight of each sample.

FIGURE 7.3 Incubation of fine root litter bags in the field.

The yearly decomposition rate (RD) for each rubber stand has been accounted from the value derived by subtracting the residual root mass recorded on final sampling date (RD_f) from the initial root mass (RD_i). The followed expression was,

$$RD = RD_i - RD_f.$$

Fine root decomposition rate constant *(k)* under different ages of rubber tree plantations were estimated using the following exponential formula (Wieder and Lang, 1982):

$$M_t/M_0 = e^{-kt},$$

where M_0 is the initial dry mass of the fine root, M_t is the remaining fine root dry mass at time *t* and *k* is the decomposition rate constant in year^{-1}.

Mean resident time of fine roots under different ages of rubber tree plantation was calculated using the following formula (Giardina and Ryan, 2000):

$$MRT = 1/k.$$

Since root mass loss due to decomposition is related to the dead root pools (Hertel and Leuschner, 2002), here in this contribution we also estimated the loss of root mass (Mg) for every month due to decomposition. The summed up values of monthly root mass loss was considered as the yearly root mass loss for the respective rubber stand. The followed calculation is as follows:

Estimated loss of root mass in ith month, EM_{loss} (Mg) = Estimated dead root biomass, ERB_{dead} of ith−1 month (Mg) × Fraction of root mass loss in ith month (%),

where $i = 1, 2, 3, \ldots, n$.

Finally, the FRP of the year for different ages of rubber plantations were estimated by using the following equation,

Fine root production of the year, FRP (Mg ha^{-1}) = Estimated live root biomass during the year, ERB_{live} (Mg ha^{-1}) + Estimated dead root biomass during the year, ERB_{dead} (Mg ha^{-1}) + Estimated mass loss during the year (Mg ha^{-1}).

Carbon stock in fine roots under different age of rubber plantations was estimated using the biomass carbon fraction proposed by Liu et al. (2017). The release of carbon from the fine root through decomposition has been estimated using the value of mass loss (Mg ha^{-1}). The average release rate of fine root carbon to the soil under rubber plantation has been estimated by averaging all the values of mass loss carbon estimated for all the stands.

7.2.2 Fine Root Biomass

The monthly observed live root biomass over the sampling period under 6-year-old rubber tree plantation ranged from 0.44 to 1.53 Mg ha^{-1}, whereas under 34-year-old plantation it was 0.86 to 3.30 Mg ha^{-1} (Fig. 7.4). The observed monthly dead root biomass over the sampling period under 6- and 34-year-old rubber plantations ranged from 0.10 to 0.44 Mg ha^{-1} and 0.35 to 1.16 Mg ha^{-1}, respectively (Fig. 7.4). However, the minimum share of live root to the total standing fine root biomass was 41% under 15-year-old stand and maximum was 89% under 27-year-old stand (Fig. 7.5). The cumulative monthly rainfall during the sampling period ranged from 3.4 to 1026.1 mm in January and August, respectively (Fig. 7.1).

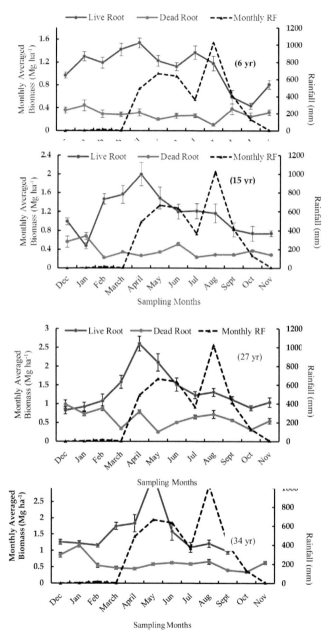

FIGURE 7.4 (See color insert.) Monthly observed root biomass with cumulative rainfall (RF) under rubber plantations of Barak Valley.

Fine Root Dynamics and Carbon Management

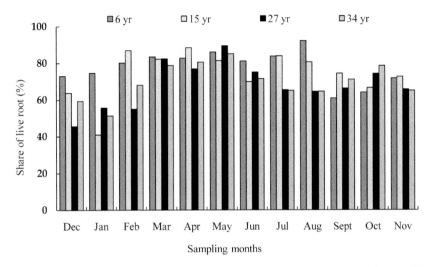

FIGURE 7.5 (See color insert.) Monthly share of live root biomass to the monthly standing fine root biomass.

The correlation coefficient (r) values showed the positive relationship of live root with rainfall but the relationship was negative for dead root biomass with rainfall (Table 7.1). The negative relationship between live root and dead root biomass was also observed in all the rubber plantations (Table 7.1). The estimated live fine root biomass ranged from 1.31 to 2.43 Mg ha^{-1} for 6- to 34-year-old stand, respectively, whereas for dead root it ranged from 0.55 to 0.94 Mg ha^{-1} for the same respective stands (Table 7.2; Fig. 7.6).

TABLE 7.1 Correlation Coefficient Values of Different Relationships at 95% Confidence Level.

	Monthly RF (mm)	Dead root 6 years (Mg ha^{-1})	Dead root 15 years (Mg ha^{-1})	Dead root 27 years (Mg ha^{-1})	Dead root 34 years (Mg ha^{-1})
Monthly RF (mm)	1.000	−0.701	−0.246	−0.151	−0.197
Live root 6 years (Mg ha^{-1})	0.169	−0.077	–	–	–
Live root 15 years (Mg ha^{-1})	0.313	–	−0.491	–	–

TABLE 7.1 *(Continued)*

	Monthly RF (mm)	Dead root 6 years (Mg ha⁻¹)	Dead root 15 years (Mg ha⁻¹)	Dead root 27 years (Mg ha⁻¹)	Dead root 34 years (Mg ha⁻¹)
Live root 27 years (Mg ha⁻¹)	0.481	–	–	−0.213	–
Live root 34 years (Mg ha⁻¹)	0.326	–	–	–	−0.109
Mass loss 6 years (%)	0.688	–	–	–	–
Mass loss 15 years (%)	0.656	–	–	–	–
Mass loss 27 years (%)	0.485	–	–	–	–
Mass loss 34 years (%)	0.622	–	–	–	–

TABLE 7.2 Estimated Biomass for Different Traits of Fine Root Dynamics.

	6 years	15 years	27 years	34 years
Live root (Mg ha⁻¹)	1.31 (0.05)	1.66 (0.20)	2.06 (0.23)	2.43 (0.08)
Dead root (Mg ha⁻¹)	0.55 (0.09)	0.68 (0.09)	0.91 (0.10)	0.94 (0.05)
Decay constant (year⁻¹)	1.04 (0.16)	1.21 (0.18)	1.24 (0.15)	1.38 (0.21)
MRT (years)	0.96	0.82	0.81	0.73
Mass loss (Mg ha⁻¹)	0.23	0.34	0.52	0.58
Fine root production (Mg ha⁻¹)	2.10	2.67	3.49	3.95

Fine Root Dynamics and Carbon Management 131

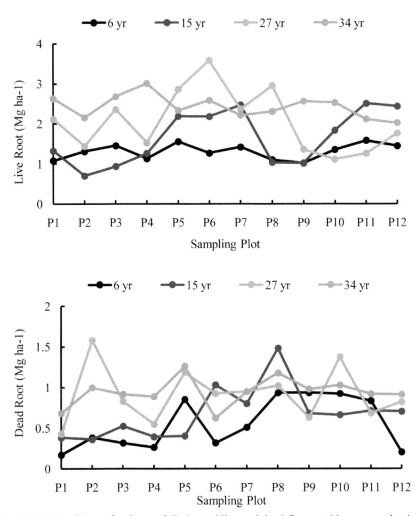

FIGURE 7.6 (See color insert.) Estimated live and dead fine root biomass production under different ages of rubber plantations.

Fine root biomass is known to be influenced by the genetic variations, soil characteristics, species variety, and seasonal variations (Jamro et al., 2015; Di Iorio et al., 2013). In the present study, monthly observed fine root biomass has been evolved to be influenced by stand age and seasonal variations. The monthly observed live root biomass has increased with the age of the stand and it ranged from 0.44 to 3.30 Mg ha^{-1} under 6- and

34-year-old plantations, respectively. Fine root biomass under rubber plantations is affected by the stand age (Liu et al., 2017). Results from northeast India showed, the highest live root under rubber plantations on the month of April (6, 15, and 27 years) and May (34 years). However, in a specific environment, age of the stand may be an influential factor of FRP (Liu et al., 2017). The monthly observed live root biomass during the high rainfall months (June, July, and August) was not affected by the stand age in northeast India. The highest peak in fine root biomass was observed under different plantations (*Fokienia hodginsii, Ormosia xylocarpa,* and *Castanopsis kawakamii)* of China in the month of March (Yang et al., 2004). However, the similar results for minimum fine root biomass were observed under *Cunninghamialanceolat, Fokienia hodginsii, Ormosia xylocarpa,* and *Castanopsis kawakamii* plantations (Yang et al., 2004). Dry period possesses all the observations of minimum live fine root biomass in the study from northeast India. Proliferation of fine roots after the first rain has been reported in plant science (Chairungsee et al., 2013). In northeast India, the dry season (RF < 50 mm in a month) ends up on the month of March and starts raining April onward result into the highest fine root proliferation under all the plantation ages. In addition to that, the monthly observed dead root biomass also increased with the increase in the age of the plantation. Results also revealed that, for a specific month, the observed dead root biomass values were significantly different among the stand ages ($P < 0.05$). The highest dead root biomass was observed on the month of January (6, 15, and 34 years) and December (27 years). The highest observation of dead root biomass during the dry season has also been supported by the emergent research findings of fine root dynamics (Assefa et al., 2017; Violita et al., 2016). Over the sampling period, the monthly fine root biomass in sequential coring method was largely shared by the live roots. The maximum share of live root was 85%–89% and was observed in the month of April and May. Significant proliferation of fine root after first rain increases the allocation of live root (Mulkey et al., 2012). Therefore, due to lack of soil moisture during dry seasons live root proliferations may get impaired significantly (Chairungsee et al., 2013). Root production in northeast India was estimated following Max–Min method. The total value for live and dead root biomass production ranged from 1.86 to 3.37 Mg ha^{-1} for 6- and 34-year-old rubber plantations, respectively. The FRP under rubber plantations of Brazil and Thailand were a little higher than that observed in northeast India (Maeght et al.,

2015; Wauters et al., 2008). However, the fine root biomass under rubber tree plantations has been estimated to be higher than that of under tropical rainforest (1.7 Mg ha^{-1}; Green et al., 2005).

7.3 FINE ROOT DECOMPOSITION

The estimated monthly mass loss due to decomposition was ranged from 3% to 13% for 6 year and 2 to 17% for 34-year-old rubber tree plantations respectively (Fig. 7.7). However, monthly mass losses under different ages of rubber plantations were positively correlated with the monthly rainfall (Table 7.1; Fig. 7.8).

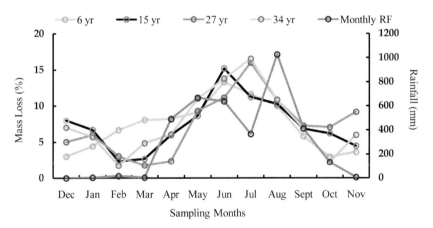

FIGURE 7.7 (See color insert.) Estimated mass loss due to decomposition in different months over the study period.

The yearly mass loss due to decomposition ranged from 0.23 to 0.58 Mg ha^{-1} under 6- and 34-year-old rubber plantations, respectively (Table 7.2). The estimated decay constant (k) for fine roots of rubber plantation was ranged from 1.04 and 1.38 for under 6- and 34-year-old plantations, respectively (Table 7.2). The mean resident time of the fine roots ranged from 0.73 for 34 year to 0.96 for 6-year-old rubber plantations (Table 7.2). The FRP under rubber tree plantations was estimated at 2.10–3.95 Mg ha^{-1} (Table 7.2). In northeast India, root mass remaining after 1 year of incubation ranged from 7% to 13% of root mass under 34- and

6-year-old plantations, respectively. The highest percentage of mass loss under different ages of rubber tree plantations was observed in June and July. Adequate soil moisture and nutrients may boost up the micro-organic activities that may lead the result to the higher decomposition during wet seasons (Sierra et al., 2015). However, lack of soil moisture due to lack of rainfall may lead to the lower micro-organic activities (Sierra et al., 2015), resulted in lowest mass loss during the month of February and March in northeast India. In addition to that, the results also revealed the positive correlations between mass loss and monthly rainfall ($P < 0.05$). Root disappearance is an age-dependent factor with average rainfall (Green et al., 2005). The yearly mass loss due to decomposition has also been estimated to increase with the age of stand and it ranged from 0.23 to 0.58 Mg ha^{-1} under 6-and 34-year-old plantations, respectively. The estimated fine root decay constant (k) under rubber tree plantations increased from 1.04 to 1.38 under 6- and 34-year-old, respectively. However, one way analysis of variance (ANOVA) reveal no significant differences between decay constant for different ages of rubber plantations ($P > 0.05$; $F = 0.604$; $P = 0.616$). Increase of decay constant is corresponding to decrease of mean resident time. Therefore, the consecutive decrease of mean residence time with the increase of plantation age has been revealed in the present study. The mean residence time (years) of fine root of rubber plantations of 0.96 year was decreased to 0.73 year for 6- and 34-year-old stands, respectively.

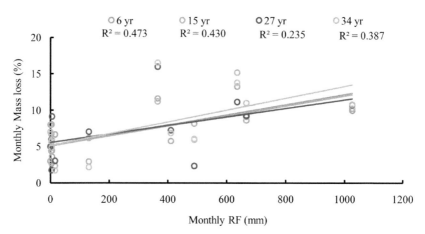

FIGURE 7.8 (**See color insert.**) Coefficient of determination for the relationship between monthly mass loss and monthly rainfall.

FRP (Mg ha^{-1}) of a system is the cumulative measures of live root production, dead root production, and decomposed root mass or mass loss over the period. FRO may vary with the age of the stand, as it possesses significant effect on live root production, dead root production, and mass loss. In northeast India, the FRP ranged from 2.10 to 3.95 Mg ha^{-1} under 6- and 34-year-old plantations, respectively. Comparative results for fine roots under 7–25-year-old rubber plantations were reported from Brazil (Liu et al., 2017). The average increment of FRP from 6- to 34-year-old rubber plantations has been estimated as 0.07 Mg ha^{-1} year^{-1}.

7.4 FINE ROOT CARBON STOCK

Carbon sink capacity under different ages of rubber plantations has been estimated at 1.01 to 1.90 Mg ha^{-1} year^{-1}. The range of estimated carbon sink capacity into live and dead root was 0.63 to 1.17 Mg ha^{-1} and 0.26 to 0.45 Mg ha^{-1}, respectively. Carbon release to the soil from fine root through decomposition process was ranged from 0.11 to 0.28 Mg ha^{-1} (Table 7.3). Shapiro–Wilk test reveal that the observed monthly live root biomass (MOLR), dead root biomass (MODR) and estimated live root biomass productions (ELRBP) for rubber plantations significantly deviates from normal distribution ($P < 0.05$), whereas estimated dead root biomass production (EDRBP) for rubber plantations were found to follow the normality trend ($P > 0.05$; Table 7.4).

TABLE 7.3 Carbon Stock for Live Root Production (LR), Dead Root Production (DR) and Total Fine Root Production (FRP), Mass Loss and Carbon Release Rate to the Soil Through Decomposition (CRD).

	6 years	15 years	27 years	34 years
LR (Mg ha^{-1})	0.6288	0.7968	0.9888	1.1664
DR (Mg ha^{-1})	0.264	0.3264	0.4368	0.4512
FRP (Mg ha^{-1} year^{-1})	1.008	1.2816	1.6752	1.896
CRD (Mg ha^{-1} year^{-1})	0.1104	0.1632	0.2496	0.2784

TABLE 7.4 Test of Normality Table for Different Traits.

Traits	Stand age (years)	Shapiro–Wilk statistics	df	Sig.
MOLR	6	0.956	144	0.000
	15	0.928	144	0.000
	27	0.902	144	0.000
	34	0.785	144	0.000
MODR	6	0.842	144	0.000
	15	0.685	144	0.000
	27	0.943	144	0.000
	34	0.930	144	0.000
ELRBP		0.950	48	0.040
EDRBP		0.973	48	0.317

Note: Table contains the results of Shapiro–Wilk test of normality (at 95% confidence level) for MOLR, MODR, ELRBP, and EDRBP of different ages of rubber plantations of Barak Valley, Assam.

ELRBP, estimated live root biomass production; EDRBP, estimated dead root biomass production; MODR, monthly observed dead root biomass; MOLR, monthly observed live root biomass.

The Kruskal–Wallis test revealed, under a specific age of rubber plantation the monthly observed live root (MOLR) and monthly observed dead root (MODR) biomass values were significantly different for various months ($P < 0.05$; Table 7.5). However, the effect of age on monthly observed live root (MOLR) and dead root biomass (MODR) was found significantly different for almost all of the studied months ($P < 0.05$; Table 7.6).

TABLE 7.5 Monthly Variations of Averaged Live Root and Dead Root Biomass for Rubber Plantation in Barak Valley, Assam.

Traits	Stand age (years)	Chi-square	df	Asymp. (sig.)
MOLR	6	72.194	11	0.000
	15	51.584	11	0.000
	27	68.412	11	0.000
	34	63.973	11	0.000
MODR	6	29.036	11	0.002
	15	99.000	11	0.000
	27	88.700	11	0.000
	34	76.325	11	0.000

Note: Table contains the results of Kruskal–Wallis test (at 95% confidence level) for the monthly variations on the MOLR and MODR under each ages of rubber plantations of Barak Valley, Assam.

MOLR, monthly observed live root biomass; MODR, monthly observed dead root biomass.

TABLE 7.6 Age Variations on Averaged Live Root and Dead Root Biomass Over the Sampling Period.

Traits	Month	Chi-square	df	Asymp. Sig.
MOLR	December	10.761	3	0.013
	January	21.539	3	0.000
	February	12.426	3	0.006
	March	11.185	3	0.011
	April	16.123	3	0.001
	May	30.875	3	0.000
	June	3.765	3	0.288
	July	1.554	3	0.670
	August	2.704	3	0.440
	September	10.608	3	0.014
	October	18.563	3	0.000
	November	11.666	3	0.009
MODR	December	24.034	3	0.000
	January	24.442	3	0.000
	February	27.971	3	0.000
	March	10.222	3	0.017
	April	28.375	3	0.000
	May	31.571	3	0.000
	June	21.243	3	0.000
	July	32.804	3	0.000
	August	39.558	3	0.000
	September	18.202	3	0.000
	October	10.210	3	0.017
	November	19.518	3	0.000
ELRBP		17.963	3	0.000
EDRBP		4.850	47	0.005

Note: Table contains the results of Kruskal–Wallis test (at 95% confidence level) for the age variations on the MOLR, MODR, and ELRBP. One-way ANOVA was applied to reveal the significant variations on EDRBP over the study period.

EDRBP, estimated dead root biomass production; ELRBP, estimated live root biomass production; MODR, observed monthly dead root biomass; MOLR, monthly observed live root biomass.

The total carbon stock produced in 1 year through the FRP under rubber plantations ranged from 1.01 to 1.9 Mg ha^{-1}. Findings from northeast India resembles with the rubber fine root carbon stock in China (1.18 to 1.99 Mg ha^{-1}; Liu et al., 2017). Increase of carbon accumulation in fine root of rubber plantations with the stand age has also been reported from China (Liu et al., 2017). The progressive increment of total carbon stock from 6- to 15-, 16- to 27-, 27- to 34-year-old rubber plantations has been estimated as 27%, 31%, and 13% of the base stand. However, the average net increment from 6- to 34-year-old rubber plantation has been estimated at 0.03 Mg ha^{-1} year^{-1}. Carbon input into the soil through fine root decomposition ranged from 0.11 to 0.28 Mg ha^{-1} year^{-1}. The average rate of carbon release from the fine root biomass to the soil is 0.20 Mg ha^{-1} yea`r^{-1}.

7.5 CONCLUSIONS

The following conclusions can be drawn from the above discussion; (1) the FRP under rubber tree plantations increase with the stand age, (2) FRP is higher during wet season than that of during dry season, (3) fine root decomposition rate or decay constant increase with the stand age, (4) mean residence time of fine roots decrease with the stand age under rubber plantation, and (5) carbon translocation rate from fine root to the soil is 0.20 Mg ha^{-1} under rubber plantations.

KEYWORDS

- **root dynamics**
- **belowground biomass**
- **root decomposition**
- **decay constant**
- **residual roots**

REFERENCE

Aber, J. D.; Melillo, J. M.; Nadelhoffer, K. J.; McClaugherty, C. A.; Pastor J. Fine Root Turnover in Forest Ecosystems in Relation to Quantity and Form of Nitrogen Availability: A Comparison of Two Methods *Oecologia* **1985**, *66*, 317–321.

Al Afas, N.; Marron, N.; Zavalloni, C.; Ceulemans, R. Growth and Production of a Short-Rotation Coppice Culture of Poplardiv: Fine Root Characteristics of Five Poplar Clones. *Biomass Bioenergy* **2008**, *32*, 494–502.

Assefa, D.; Rewald, B.; Sandén, H.; Godbold, D. L. Fine Root Dynamics in Afromontane Forest and Adjacent Land Uses in the Northwest Ethiopian Highlands. *Forests* **2017**, *8* (7), 249.

Bennett, J. N.; Andrew, B.; Prescott C. E. Vertical Fine Root Distributions of Western Redcedar, Western Hemlock, and Salal in Old-Growth Cedar-Hemlock Forests on Northern Vancouver Island. *Can. J. For. Res.* **2002**, *32*, 1208–1216.

Bloom, A. A.; Exbrayat, J. F.; Velde, I. R. V. D.; Feng, L.; Williams, M. The Decadal State of the Terrestrial Carbon Cycle: Global Retrievals of Terrestrial Carbon Allocation, Pools, and Residence Times. *PNAS* **2016**, *113*, 1285–1290.

Brahma, B.; Nath, A. J.; Das, A. K. Managing Rubber Plantations for Advancing Climate Change Mitigation Strategy. *Curr. Sci.* **2016**, *110*, 2015–2019.

Brahma, B.; Sileshi, G. W.; Nath, A. J.; Das, A. K. Development and Evaluation of Robust Tree Biomass Equations for Rubber Tree (*Hevea brasiliensis*) Plantations in India. *For. Ecosyst.* **2017**, *4*, 1–14. DOI: 10.1186/s40663-017-0101-3

Brassard, B. W.; Chen, H. Y.; Bergeron, Y.; Pare, D. Differences in Fine Root Productivity Between Mixed and Single Species Stands. *Funct. Ecol.* **2011**, *25*, 238–246.

Chairungsee, N.; Gay, F.; Thaler, P.; Kasemsap, P.; Thanisawanyangkura, S.; Chantuma, A. et al. Impact of Tapping and Soil Water Status on Fine Root Dynamics in a Rubber Tree Plantation in Thailand. *Front. Plant Sci.* **2013**, *4*, 538.

Chimento, C.; Amaducci, S. Characterization of Fine Root System and Potential Contribution to Soil Organic Carbon of Six Perennial Bioenergy Crops. *Biomass Bioener.* **2015**, *83*, 116–122.

De Graaff, M. A.; Six, J.; Jastrow, J. D.; Schadt, C. W.; Wullschleger, S.D. Variation in Root Architecture Among Switchgrass Cultivars Impacts Root Decomposition Rates. *Soil Biol. Biochem.* **2013**, *58*, 198–206.

Fahey, T. J.; Hughes, J. W.; Mou, P.; Arthur, M. A. Root Decomposition and Nutrient Flux Following Whole-Tree Harvest of Northern Hardwood Forest. *For. Sci.* **1988**, *34*, 744–768.

Finer, L.; Helmisaari, H. S.; Lõhmus, K.; Majdi, H.; Brunner, I.; Børja, I. et al. Variation in Fine Root Biomass of Three European Tree Species: Beech (*Fagussylvatica* L.), Norway Spruce (*Piceaabies* L. Karst.), and Scots Pine (*Pinussylvestris* L.). *Plant Biosyst.* **2007**, *141* (3), 394–405.

Giardina, C. P.; Ryan, M. G. Evidence that Decomposition Rates of Organic Carbon in Mineral Soil do not Vary with Temperature. *Nature* **2000**, *404*, 858–861.

Gower, S. T.; Pongracic, S.; Landsberg, J. J. A Global Trend in Belowground Carbon Allocation: Can We Use the Relationship at Smaller Scale? *Ecology* **2014**, 77, 1750–1755.

Green, J. J.; Dawson, L. A.; Proctor, J.; Duff, E. I.; Elston, D. A. Fine Root Dynamics in a Tropical Rain Forest Is Influenced by Rainfall. *Plant Soil* **2005**, *276*, 23–32.

Hendricks, J. J.; Hendrick, R. L.; Wilson, C. A.; Mitchell, R. J.; Pecot, S. D.; Guo, D. Assessing the Patterns and Controls of Fine Root Dynamics: An Empirical Test and Methodological Review. *J. Ecol.* **2006**, *94*, 40–57.

Hertel, D.; Leuschner, C. A. Comparison of Four Different Fine Root Production Estimates with Ecosystem Carbon Balance Data in a Fagus-Quercus Mixed Forest. *Plant Soil* **2002**, *239*, 237–251.

Di Iorio, A.; Montagnoli, A.; Terzaghi, M.; Scippa, G. S.; Chiatante, D. Effect of Tree Density on Root Distribution in *Fagussylvatica* Stands: A Semi-Automatic Digitising Device Approach to Trench Wall Method. *Trees* **2013**, *27*, 1503–1513.

Jackson, R. B.; Mooney, H. A.; Schulze, E. D. A Global Budget for Fine Root Biomass, Surface Area, and Nutrient Contents. *PNAS* **1997**, *94*, 7362–7366.

Jamro, G. M.; Chang, S. X.; Naeth, M. A.; Duan, M.; House, J. Fine Root Dynamics in Lodgepole Pine and White Spruce Stands Along Productivity Gradients in Reclaimed Oil Sands Sites. *Ecol. Evol.* **2015**, *5* (20), 4655–4670.

Jessy, M. D.; Prasannakumari, P.; Abraham, J. Carbon and Nutrient Cycling Through Fine Roots in Rubber (*Hevea brasiliensis*) Plantations in India. *Exper. Agri.* **2013**, *49*, 556–573.

Kong, A. Y. Y.; Six, J. Tracing Root vs. Residue Carbon into Soils from Conventional and Alternative Cropping Systems. *Soil Biol. Biochem.* **2010**, *74*, 1201–1210.

Lal, R. Soil Carbon Sequestration Impacts on Global Climate Change and Food Security. *Science* **2004**, *304* (5677), 1623–1627.

Lehmann, J.; Zech, W. Fine Root Turnover of Irrigated Hedgerow Intercropping in Northern Kenya. *Plant Soil.* **1998**, *198* (1), 19–31.

Leuschner, C.; Hertel, D.; Schmid, I.; Koch, O.; Muhs, A.; Hölscher, D. Stand Fine Root Biomass and Fine Root Morphology in Old-Growth Beech Forests as a Function of Precipitation and Soil Fertility. *Plant Soil* **2004**, *258* (1), 43–56.

Liu, C.; Pang, J.; Jepsen, M. R.; Lü, X.; Tang, J. Carbon Stocks Across a Fifty Year Chronosequence of Rubber Plantations in Tropical China. *Forests* **2017**, *8*, 209.

Maeght, J. L.; Gonkhamdee, S.; Clément C.; Na Ayutthaya, S.; Stokes, A.; Pierret, A. Seasonal Patterns of Fine Root Production and Turnover in a Mature Rubber Tree (*Hevea brasiliensis*Müll.Arg.) Stand Differentiation with Soil Depth and Implications for Soil Carbon Stocks. *Front. Plant Sci.* **2015**, *6*, 1022.

McClaugherty, C. A.; Aber, J. D.; Melillo, J. M. The Role of Fine Roots in the Organic Matter and Nitrogen Budgets of Two Forested Ecosystems. *Ecology* **1982**, *63* (5), 1481–1490.

McClaugherty, C. A.; Aber, J. D.; Melillo, J. M. Decomposition Dynamics of Fine Roots in Forested Ecosystems. *Oikos* **1984**, 378–386.

Mulkey, S. S.; Chazdon, R. L.; Smith, A. P. *Tropical Forest Plant Ecophysiology*, 1st ed.; Springer: US, 1996.

Olmo, M.; Lopez-Iglesias, B.; Villar, R. Drought Changes the Structure and Elemental Composition of Very Fine Roots in Seedlings of Ten Woody Tree Species, Implications for a Drier Climate. *Plant Soil* **2014**, *384*, 113–129.

Publicover, D. A.; Vogt, K. A. A Comparison of Methods for Estimating Forest Fine Root Production with Respect to Sources of Error. *Can. J. For. Res.* **1993**, *23*, 1179–1186.

Puget, P.; Drinkwater, L.E. Short-term Dynamics of Root- And Shoot-Derived Carbon from a Leguminous Green Manure. *Soil Sci. Soc. Am. J.* **2001**, *65*, 771–779.

Rasse, D. P.; Rumpel, C.; Dignac, M. F. Is Soil Carbon Mostly Root Carbon? Mechanisms for a Specific Stabilisation. *Plant Soil.* **2005,** *269,* 341–356.

Ruess, R. W.; Van, C. K.; Yarie, J.; Viereck, L. A. Contributions of Fine Root Production and Turnover to the Carbon and Nitrogen Cycling in Taiga Forests of the Alaskan Interior. *Can. J. For. Res.* **1996,** *26,* 1326–1336.

Sierra, C. A.; Trumbore, S. E.; Davidson, E. A.; Vicca, S.; Janssens, I. Sensitivity of Decomposition Rates of Soil Organic Matter with Respect to Simultaneous Changes in Temperature and Moisture. *J. Adv. Mod. Earth Syst.* **2015,** *7,* 335–356.

Thaler, P. Root Apical Diameter and Root Elongation Rate of Rubber Seedlings (*Hevea brasiliensis*) Show Parallel Responses to Photo Assimilate Availability. *Physiologia Plantarum* **1996,** *97,* 365–71.

UNFCCC. Report of the Conference of the Parties on Its Thirteenth Session, Bali, December 3–15, 2007, Addendum Part Two: Action Taken by the Conference of the Parties at Its Thirteenth Session; 2008.

Violita, V.; Triadiati, T.; Anas, I.; Miftahudin, M. Fine Root Production and Decomposition in Lowland Rainforest and Oil Palm Plantations in Sumatra, Indonesia. *HAYATI J. Biosci.* **2016,** *23,* 7–12.

Vogt, K. A.; Vogt, D. J.; Palmiotto, P. A.; Boon, P.; O'Hara, J.; Asbjornsen, H. Review of Root Dynamics in Forest Ecosystems Grouped by Climate, Climatic Forest Type and Species. *Plant Soil.* **1996,** *187,* 159–219.

Wauters, J. B.; Coudert, S.; Grallien, E.; Jonard, M.; Ponette, Q. Carbon Stock in Rubber Tree Plantations in Western Ghana and MatoGrosso (Brazil). *For. Ecol. Manag.* **2008,** *255,* 2347–2361.

Wei, X. R.; Shao, M. G.; Gale, W.; Li, L. H. Global Pattern of Soil Carbon Losses Due to the Conversion of Forests to Agricultural Land. *Sci. Rep.* **2014,** *4,* 4062.

Wieder, R. K.; Lang, G. E. A Critique of the Analytical Methods Used in Examining Decomposition Data Obtained from Litter Bags. *Ecology* **1982,** *63,* 1636–1642.

Yang, Y.; Chen, G.; Lin, P.; Xie, J.; Guo, J. Fine Root Distribution, Seasonal Pattern and Production in Four Plantations Compared with a Natural Forest in Subtropical China. *Ann. For. Sci.* **2004,** *61,* 617–627.

CHAPTER 8

Conclusions and Recommendations

Rubber, an important agricultural plantation in tropics, is cultivated for natural latex production in all tropical zones, specifically in the 10° either latitudinal side of the equator. It is rapidly expanding into both climatically optimal and sub-optimal environments in almost all rubber-growing countries across the world. The species prefers an annual mean temperature range of 20–30°C, precipitation >1500 mm, light to heavy clay with good water drainage for its growth. Attracted by the economic benefits and government incentives to convert traditional farming areas into high-value commercial crops, many farmers switched to rubber cultivation in Asia. Consequently, large areas of natural forests, degraded forests and other crops have been cleared to grow rubber plantations, which has emerged as the most widespread smallholder tree crop. One of the major factors driving transition to more intensive agriculture, monocropping, and crop replacement has been population growth, including internal migration. Besides, there are infrastructural developments like expansion of road network and markets making it easier for farmers to purchase agricultural inputs and to sell their crops. Government policy and incentives have also given a shift towards rubber plantation.

The conversion of natural forest to rubber plantations has often been associated with various environmental issues including biodiversity loss, disruption of hydrological processes and nutrient cycles, and soil carbon loss. There is widespread concern about environmental impact of rubber plantations among the rubber producing countries. Biodiversity loss may also lead to reduced total carbon biomass thereby impacting climate change. Rubber plantations are often cultivated between rows in the early years and this practice produces significant erosion on slopes. The case of land use conversion of natural forests into rubber plantation and associated environmental issues in the mainland South East Asia, where rubber plantation covers five lakh hectares at present and by 2050 it may be tripled.

Another study in Southeast Asia indicated greater annual catchment water losses through evapotranspiration from the rubber dominated catchments compared to traditional vegetation cover. Drought in many parts of the world has been related with the rubber plantation expansion. For example, Tripura, state of India which had always been a water surplus state, is ironically suffering from drought. Tripura happens to be the second highest grower of rubber in India and the problems of water have only surfaced over the last 5 years. Similarly, Indonesia once the highest cultivator of commercial rubber has not only banned the tree but has also burned down whole jungles of rubber plantations. While economically this was done due to a fall in the market price, but it has been commonly known that the advent of rubber led to a dip in the water levels of the country. Therefore alternatively, rubber agroforestry could be a strategy to combat consequences of environmental problems. Elsewhere rubber agroforestry systems have been known to support relatively higher biodiversity, greater C stocks, and local livelihoods although they may have relatively low latex productivity compared to monoculture plantations. Promotion of diversified agroforestry systems in which rubber plays an important role, but is not planted as monocultures may be more beneficial. Intercropping of banana has brought good results in experimental studies in Sri Lanka.

Use of tree diameter at 1.37 m as an independent variable for estimation of biomass is well recognized. Rubber trees are managed for latex collection through tapping of stems up to 2 m tree height from base. Therefore, the effect of tapping in the stem can blunt the accuracy of biomass model. To avoid these art effects, diameter at 2 m height of the tree may be considered as the variable for developing biomass model. Incorporating height of the tree with the diameter may eliminate the ambiguity of the developed model. Since the biomass of the tree varies with the age, the development of biomass model for rubber trees incorporating compound variable of diameter, height, and age of the tree can yield the best predictive model.

Despite their negative environmental impacts mentioned above, rubber plantations have immense potential for terrestrial carbon sequestration by accumulating carbon in woody biomass and organic matter in mineral soil. On a global scale biomass carbon stock (86–170 Mg ha^{-1}) of mature rubber plantations are comparable with that under many forests and agroforestry systems of the world. Therefore, management of rubber plantations can play an important role in mitigating climate change. In rubber plantations, the proportion of recalcitrant carbon pools increase with increase in age of the rubber plantation indicating the potentiality of rubber plantations in long term carbon sink management. In a comparative study it was observed that the SOC stocks under 34-year-old plantations were 20% higher than that under *Imperata cylindrica* grassland, but 34% lower than SOC stocks recorded under natural forest soil. Therefore, rubber plantation development by replacing natural forest should not be encouraged; however, *Imperata* grasslands may be converted to rubber plantations for enhanced biomass stock management. Therefore, expansion of rubber tree plantations with best management practices in degraded forest, grassland or fallow lands can offer an ecological stability with significant socioeconomic support.

KEYWORDS

- **biodiversity loss**
- **rubber agroforestry**
- **environmental issues**
- **local livelihood**
- **monoculture**

Index

A

Aboveground biomass (AGB)
  age-specific models of, 28
  fitted lines of the models used, 32
Allometric model, 50
Allometric scaling relationships, 51
Annual mean temperature, 143

B

Barak Valley (BV), 85
Biodiversity loss, 143
Biomass carbon
  C storage and subsequent loss
    comparison of, 53
    mean and median, 52
  data collection, methods
    C densities, estimation of, 47–49
    stand selection and sampling, 46–47
    tree biomass, estimation of, 47–49
  estimation models
    allometric model, 50
    allometric scaling relationships, 51
    site specific models and general model, 50
    strong allometric relationships, 49
    total biomass, 50
  potential loss of
    C sequestration, 45
    conversion of forests, 44
    destructive and non-destructive methods, 45
    factors, 43
    land-use change, risk, 44
    plantations of, 43
    rubber trees, 44
  potential stocks loss
    AGB and BGB, 53
    allometric relationships, 53
    clear-felling, 56
    comparison of models, 54
    conversion of forests, 56
    felling cycle, 55
    latex production, 55
    mature rubber plantations, 55
    rubber agroforestry systems, 55–56
    total biomass of, 54
    traditional practice, 53
  stocks loss, 19
    AGB and BGB, 53
    allometric relationships, 53
    clear-felling, 56
    comparison of models, 54
    conversion of forests, 56
    felling cycle, 55
    latex production, 55
    mature rubber plantations, 55
    rubber agroforestry systems, 55–56
    total biomass of, 54
    traditional practice, 53
  tree biomass
    comparison of, 53
    mean and median, 52
Biomass models
  aboveground biomass (AGB)
    age-specific models of, 28
    fitted lines of the models used, 32
  carbon stock in rubber plantations
    global status of, 36–38
  clean development mechanism (CDM), 20
  cumulative anthropogenic, 19
  data generation, methods for
    diameter at breast height (DBH), 22
    part dry weight (PDW), 22
    ten plots, 22
  development
    diameter, 23
    exponential, 24
    Gompertz, 24

height–diameter (H–D) models, 24
Hyperbolic models, 24
logistic, 24
measured tree height, absence, 24
monomolecular, 24
multiplicative error structure, 24
performance of, 25
power-law, 24
power-law scaling, 23
Richards function, 24
root dry weight (RDW), 23
stand density (StD), 23
Weibull, 24–25
estimation
  aboveground biomass (AGB), 26
  AIC and RMSE, 29
  average tree density, 26
  coarse root biomass, 26, 33
  foliage dry weight, 33
  H–D models, 27
  height (cm), 26
  mean absolute percentage error (MAPE), 33
  and model fit criteria, 26
  parameter estimates, 26
  parameters and goodness of fit statistics for, 29
  power-law model, 30
  relationship between, 27
  stem diameter at 200 cm, 26
  strong non-linear relationship, 26
  total biomass stocks, 26
  total dry weight, 33
  tree density, 25
Food and Agriculture Organization (FAO), 20
intergovernmental panel, 19
measured biomass
  carbon stock and sequestration rate, 35
root dry weight (RDW)
  fitted lines of, 34
total biomass stock
  destructive and non-destructive methods, 20
  destructive method, 21
  estimation of, 20
  sub-tropical trees, 21
variation in
  measured and estimated tree biomass components with, 31
  tree biomass components, 30

C

C sequestration, 45
Carbon sink capacity
  age variations, 137
  Barak Valley (BV), rubber plantation in, 136
  ELRBP, 136
  FRP under, 138
  monthly observed dead root (MODR), 136
  monthly observed live root (MOLR), 136
  progressive increment, 138
  Shapiro–Wilk test, 135
  test of normality table, 136
Carbon stock in rubber plantations
  biomass models
    global status of, 36–38
Clean development mechanism (CDM)
  biomass models, 20
Clear-felling, 56
Conversion of forests, 44, 56
Cumulative anthropogenic
  biomass models, 19

D

Data collection, methods
  ecosystem carbon sequestration
    aboveground biomass (AGB), 88
    Barak Valley (BV), 85
    biomass sampling, plot selection for, 87–89
    C sequestration rate, 89
    C stock, 85
    Imperata, 88
    ISRO-GBP/NCP-VCP protocol, 87
    Nikon Laser Forestry Pro, 88
    One way ANOVA, 90
    predominant land uses, 85–86
    and preparation, 89–90

Index

progressive and retrogressive land, schematic presentation, 87
Shapiro–Wilk tests, 90
soil sample collection, 89–90
stand selection and sampling, 46–47
tree biomass and C densities, estimation of, 47–49

Data generation
  methods for
    diameter at breast height (DBH), 22
    part dry weight (PDW), 22
    ten plots, 22

Diameter at breast height (DBH), 22

Diverse ecosystem
  land use characteristics (LUC)
    biomass stock, 91
    bulk density (BD) and, 92
    highest stem density, 90
    Levene's test, 91
    lowest and highest SOC, 90
    One-way ANOVA, 91
    Tukey HSD test, 90

E

Ecosystem carbon sequestration
  annual C emission, 83
  biomass
    decline in SOC stock, 94
    northeast India, predominant land uses of, 93–95
    progressive LUC, 93
    retrogressive LUC, 94
    RP and loss under DF, 94
    vegetation C stock, 92
  C stock
    decline in SOC stock, 94
    northeast India, predominant land uses of, 93–95
    progressive LUC, 93
    retrogressive LUC, 94
    RP and loss under DF, 94
    vegetation C stock, 92
  carbon sequestration, 96
    PB and cumulative stock, 96
    progressive LUC, 96
  combating greenhouse gas (GHG), 83

data collection, methods of
  aboveground biomass (AGB), 88
  Barak Valley (BV), 85
  biomass sampling, plot selection for, 87–89
  ecosystem C sequestration rate, 89
  ecosystem C stock, 85
  Imperata, 88
  ISRO-GBP/NCP-VCP protocol, 87
  Nikon Laser Forestry Pro, 88
  One way ANOVA, 90
  predominant land uses, 85–86
  and preparation, 89–90
  progressive and retrogressive land, schematic presentation, 87
  Shapiro–Wilk tests, 90
  soil sample collection, 89–90
loss of biodiversity, 84
LUC
  basal area, 97
  biomass C stock, changes in, 98–99
  biomass stock, 91
  bulk density (BD) and, 92
  ecosystem C stock, changes in, 99–101
  highest stem density, 90
  Imperata grass, 98
  Levene's test, 91
  lowest and highest SOC, 90
  One-way ANOVA, 91
  stem density, 97
  Tukey HSD test, 90
and management
  biomass C stock, changes in, 98–99
  ecosystem C stock, changes in, 99–101
northeast India (NEI), 84
Reduction Emission from Deforestation and Forest Degradation (REDD), 83–84
soil
  decline in SOC stock, 94
  ecosystem C sequestration, 96
  northeast India, predominant land uses of, 93–95
  PB and cumulative stock, 96
  progressive LUC, 93, 96

retrogressive LUC, 94
RP and loss under DF, 94
vegetation C stock, 92
United Nations Framework Convention on Climate Change (UNFCCC), 83
vegetation
  ecosystem C sequestration, 96
  PB and cumulative stock, 96
  progressive LUC, 96
Environmental issues, 143

F

Fine root dynamics and carbon management
  biomass
    correlation coefficient *(r)* values, 129–130
    cumulative monthly rainfall, 127
    dry period, 132
    dynamics, estimation, 130
    genetic variations, 131
    live and dead roots, total value for, 132
    northeast India, 132
    sampling period, 127
    and soil, 122
  carbon, 121
  carbon sink capacity
    age variations, 137
    Barak Valley (BV), rubber plantation in, 136
    ELRBP, 136
    FRP under, 138
    monthly observed dead root (MODR), 136
    monthly observed live root (MOLR), 136
    progressive increment, 138
    Shapiro–Wilk test, 135
    test of normality table, 136
  CO_2 emission, 122
  decomposition
    coefficient of, 134
    FRP and FRO, 135
    FRP under rubber tree plantations, 133
    mass losses, 133–134
    one way analysis of variance (ANOVA), 134
  methods
    air-dried root material, 125
    carbon stock, 127
    decomposition rate, 125–127
    exponential formula, 126
    laboratory analysis, 125
    mean resident time, 126
    nine mesh bags, 125
    root biomass (RB), 124
    root mass loss, 127
    sampling date, 126
    soil samples, 123
  natural phenomenon, 121
  production and turnover rate, 123
  rubber tree
    ecosystem carbon sink capacity, 122

H

Height–diameter (H–D) models, 24
Hevea brasiliensis. See Rubber tree plantations
Hyperbolic models, 24

L

Land use change (LUC), 61, 113
  biomass stock, 91
  bulk density (BD) and, 92
  C sequestration rate, 115
  domesticated crops, 9
  Eleven regions, 9
  evolution, 12–14
  highest stem density, 90
  Levene's test, 91
  lowest and highest SOC, 90
  and management
    ecosystem C stock, changes in, 99–101
  natural land architectures, 8
  oil palm, 114
  One-way ANOVA, 91
  primitive agriculture, 8–12, 10
  rubber plantations in, 114
  seed or plants, purposive selection, 9

Index

theory of domestication, 9
Tukey HSD test, 90
vegetative characteristics, 9
Local livelihood, 144

M

Mean absolute percentage error (MAPE), 33
Monoculture, 144

N

Natural rubber
   production in 2011, share of, 2
   Worldwide net production, 2
Northeast India (NEI), 84
   predominant land uses of, 93–95

P

Part dry weight (PDW), 22

R

Reduction Emission from Deforestation and Forest Degradation (REDD), 83–84
Root dry weight (RDW), 23
   fitted lines of, 34
Rubber agroforestry, 144
Rubber tree plantations
   Association of Natural Rubber Producing countries (ANRPC), 4
   distribution of
     synthetic rubber, 1
     in Tropical Asian countries, 1
   and expansion
     domesticated crops, 9
     Eleven regions, 9
     evolution, 12–14
     natural land architectures, 8
     primitive agriculture, 8–12, 10
     seed or plants, purposive selection, 9
     theory of domestication, 9
     vegetative characteristics, 9
   global distributions and production
     Amazon basin, 2
     China's success, 3
     climatic suitability in, 4
     Worldwide net area, 3
   land use change (LUC)
     domesticated crops, 9
     Eleven regions, 9
     evolution, 12–14
     natural land architectures, 8
     primitive agriculture, 8–12, 10
     seed or plants, purposive selection, 9
     theory of domestication, 9
     vegetative characteristics, 9
   natural forest, conversion of, 143
   natural rubber
     production in 2011, share of, 2
     Worldwide net production, 2
   tropical forest and carbon, 6
     accumulation rate and sink capacity, 6–7
     accurate estimation, 7
     active pool (AC), 8
     C stock and sequestration, 7
     forestry and agroforestry, 7
     passive pool (PC), 8
     SOC concentration, 8
     tree biomass, 7
   world carbon budget
     atmospheric concentration, 5
     carbon sinks, 6
     global facilities, 6
     greenhouse gas (GHG) emission, 5
     Kyoto Protocol (KP), 6

S

Shapiro–Wilk tests, 90
Soil carbon sequestration
   atmospheric carbon concentration, 107
   commercial rubber cultivation, 108
   concentration, effect of land use changes in
     fraction of, 112
     Tukey's HSD test, 112
   land use change (LUC), 113
     C sequestration rate, 115
     oil palm, 114
     rubber plantations in, 114
   photosynthesis, 107
   rubber plantations

NEI, 115
  rate or time-averaged, 116
  rubber tree, 108
  sampling and analyses, 109
    bulk density (BD) estimation, 110
  soil bulk density under rubber
    plantations, 111
    BD values, 111
    upper layer, 111
  statistical analysis
    honestly significant difference (HSD), 110
    post-hoc test, 110
  stock, 113
    C sequestration rate, 115
    oil palm, 114
    rubber plantations in, 114
  study site and stand characteristics
    Barak Valley (BV), 108
    natural forest, 109
    northeast India (NEI), 108
    rubber plantations, 109
  upper soil layers, 107
  vegetative and soil pools, 107
Soil organic carbon (SOC)
  active C (AC) pool, 62
  C pools, variation in
    AC and PC pools, 71–72
    fractions, 68
    labile carbon (LC), 69–70
    less labile carbon (LLC), 69–70
    NLC, proportion of, 68
    non-labile carbon (NLC), 69–70
    very labile carbon (VLC), 69–70
  conversion of forests, 61
  data collection methods
    aboveground litter production, 65
    Barak Valley (BV), 63
    belowground litter production, 65–66
    correlation analysis, 68
    grasslands in, 63
    MRT determination, 66–67
    rubber trees, 68
    soil sampling and analyses, 63–65
    soil texture, study, 63
    study area, 62–63

Tukey's honestly significant
  difference (HSD) test, 67
World Reference Base for Soil
  Resources (WRB), 63
dynamics, 61
litter production, 73
  aboveground and belowground,
    correlations of, 74
  Tukey's HSD test, 74
losses and sequestration potentials, 75
mean residence time (MRT), 62
microbial and derivative products,
  61–62
stock and lability, changes in
  aboveground and belowground, 77
  forests, conversion of, 75
  high proportion of NLC and LLC, 77
  LUC, 76
  micro-organisms, efficiency of, 76
  plantations in grassland, development, 76
  SOM in, 76
  tannin and waxy compounds, 77
  total organic carbon (TOC) stock, 62
Southeast Asia, 144
Stand density (StD), 23

T

Total biomass stock
  destructive and non-destructive
    methods, 20
  destructive method, 21
  estimation of, 20
  sub-tropical trees, 21
Tropical forest
  and carbon, 6
    accumulation rate and sink capacity,
      6–7
    accurate estimation, 7
    active pool (AC), 8
    C stock and sequestration, 7
    forestry and agroforestry, 7
    passive pool (PC), 8
    SOC concentration, 8
    tree biomass, 7

Index

U
United Nations Framework Convention on Climate Change (UNFCCC), 83

V
Vegetation
  ecosystem C sequestration, 96
  PB and cumulative stock, 96
  progressive LUC, 96

W
World carbon budget
  rubber plantations
    atmospheric concentration, 5
    carbon sinks, 6
    global facilities, 6
    greenhouse gas (GHG) emission, 5
    Kyoto Protocol (KP), 6